Toxicity of Military Smokes and Obscurants

Volume 1

Subcommittee on Military Smokes and Obscurants
Committee on Toxicology
Board on Environmental Studies and Toxicology
Commission on Life Sciences
National Research Council

NATIONAL ACADEMY PRESS
WASHINGTON, D.C., 1997

NATIONAL ACADEMY PRESS 2101 Constitution Avenue, NW Washington, DC 20418

NOTICE The project that is the subject of this report was approved by the Governing Board of the National Research Council, whose members are drawn from the councils of the National Academy of Sciences, the National Academy of Engineering, and the Institute of Medicine. The members of the committee responsible for the report were chosen for their special competences and with regard for appropriate balance.

This report has been reviewed by a group other than the authors according to procedures approved by a Report Review Committee consisting of members of the National Academy of Sciences, the National Academy of Engineering, and the Institute of Medicine.

The National Academy of Sciences is a private, nonprofit, self-perpetuating society of distinguished scholars engaged in scientific and engineering research, dedicated to the furtherance of science and technology and to their use for the general welfare. Upon the authority of the charter granted to it by the Congress in 1863, the Academy has a mandate that requires it to advise the federal government on scientific and technical matters. Dr. Bruce Alberts is president of the National Academy of Sciences.

The National Academy of Engineering was established in 1964, under the charter of the National Academy of Sciences, as a parallel organization of outstanding engineers. It is autonomous in its administration and in the selection of its members, sharing with the National Academy of Sciences the responsibility for advising the federal government. The National Academy of Engineering also sponsors engineering programs aimed at meeting national needs, encourages education and research, and recognizes the superior achievements of engineers. Dr. William A. Wulf is interim president of the National Academy of Engineering.

The Institute of Medicine was established in 1970 by the National Academy of Sciences to secure the services of eminent members of appropriate professions in the examination of policy matters pertaining to the health of the public. The Institute acts under the responsibility given to the National Academy of Sciences by its congressional charter to be an adviser to the federal government and, upon its own initiative, to identify issues of medical care, research, and education. Dr. Kenneth I. Shine is president of the Institute of Medicine.

The National Research Council was organized by the National Academy of Sciences in 1916 to associate the broad community of science and technology with the Academy's purposes of furthering knowledge and advising the federal government. Functioning in accordance with general policies determined by the Academy, the Council has become the principal operating agency of both the National Academy of Sciences and the National Academy of Engineering in providing services to the government, the public, and the scientific and engineering communities. The Council is administered jointly by both Academies and the Institute of Medicine. Dr. Bruce Alberts and Dr. William A. Wulf are chairman and vice chairman, respectively, of the National Research Council.

The project was supported by the U.S. Department of Defense under contract DAMD 17-89-C-9086.
Additional copies of this report are available from the Board on Environmental Studies and Toxicology, 2101 Constitution Avenue, NW, Washington, DC 20418.

Printed in the United States of America.

Subcommittee on the Assessment of Military Smokes and Obscurants

MICHELE A. MEDINSKY *(Chair)*, Chemical Industry Institute of Toxicology, Research Triangle Park, N.C.
DEBORAH A. CORY-SLECHTA, University of Rochester School of Medicine, Rochester, N.Y.
CHARLES E. FEIGLEY, University of South Carolina School of Public Health, Columbia, S.C.
DONALD E. GARDNER, Inhalation Toxicology Associates, Raleigh, N.C.
SIDNEY GREEN, Corning Hazelton, Inc., Vienna, Va.
ROGENE F. HENDERSON, Lovelace Biomedical and Environmental Research Institute, Albuquerque, N.Mex.
CAROLE A. KIMMEL, U.S. Environmental Protection Agency, Washington, D.C.

Staff

KULBIR S. BAKSHI, Program Director for the Committee on Toxicology
MARGARET M. MCVEY, Project Director
RUTH E. CROSSGROVE, Editor
CATHERINE M. KUBIK, Senior Program Assistant
LINDA V. LEONARD, Senior Project Assistant
LUCY V. FUSCO, Program Assistant

Sponsor: U.S. Department of Defense

Committee on Toxicology

Staff of the Committee on Toxicology

KULBIR S. BAKSHI, Program Director
MARGARET E. MCVEY, Program Officer
RUTH E. CROSSGROVE, Editor
CATHERINE M. KUBIK, Senior Program Assistant
LUCY V. FUSCO, Program Assistant
LINDA V. LEONARD, Senior Project Assistant

Board on Environmental Studies and Toxicology

Staff

Commission on Life Sciences

Other Recent Reports of the Board on Environmental Studies and Toxicology

Carcinogens and Anticarcinogens in the Human Diet: A Comparison of Naturally Occurring Synthetic and Natural Substances (1996)

Upstream: Salmon and Society in the Pacific Northwest (1996)

Science and the Endangered Species Act (1995)

Wetlands: Characteristics and Boundaries (1995)

Biologic Markers (Urinary Toxicology (1995), Immunotoxicology (1992), Environmental Neurotoxicology (1992), Pulmonary Toxicology (1989), Reproductive Toxicology (1989))

Review of EPA's Environmental Monitoring and Assessment Program (three reports, 1994-1995)

Science and Judgment in Risk Assessment (1994)

Ranking Hazardous Sites for Remedial Action (1994)

Pesticides in the Diets of Infants and Children (1993)

Issues in Risk Assessment (1993)

Setting Priorities for Land Conservation (1993)

Protecting Visibility in National Parks and Wilderness Areas (1993)

Dolphins and the Tuna Industry (1992)

Environmental Neurotoxicology (1992)

Hazardous Materials on the Public Lands (1992)

Science and the National Parks (1992)

Animals as Sentinels of Environmental Health Hazards (1991)

Assessment of the U.S. Outer Continental Shelf Environmental Studies Program, Volumes I-IV (1991-1993)

Human Exposure Assessment for Airborne Pollutants (1991)

Monitoring Human Tissues for Toxic Substances (1991)

Rethinking the Ozone Problem in Urban and Regional Air Pollution (1991)

Decline of the Sea Turtles (1990)

Tracking Toxic Substances at Industrial Facilities (1990)

Copies of these reports may be ordered from the National Academy Press (800) 624-6242 or (202) 334-3313

Other Recent Reports of the Committee on Toxicology

Toxicity of Alternatives to Chlorofluorocarbons: HFC-134a and HCFC-123 (1996)

Permissible Exposure Levels for Selected Military Fuel Vapors (1996)

Spacecraft Maximum Allowable Concentrations for Selected Airborne Contaminants, Volume I (1994), Volume 2 (1996), and Volume 3 (1996)

Nitrate and Nitrite in Drinking Water (1995)

Guidelines for Chemical Warfare Agents in Military Field Drinking Water (1995)

Review of the U.S. Naval Medical Research Institute's Toxicology Program (1994)

Health Effects of Permethrin-Impregnated Army Battle-Dress Uniforms (1994)

Health Effects of Ingested Fluoride (1993)

Guidelines for Developing Community Emergency Exposure Levels for Hazardous Substances (1993)

Guidelines for Developing Spacecraft Maximum Allowable Concentrations for Space Station Contaminants (1992)

Review of the U.S. Army Environmental Hygiene Agency Toxicology Division (1991)

Permissible Exposure Levels and Emergency Exposure Guidance Levels for Selected Airborne Contaminants (1991)

Preface

U.S. ARMY personnel are exposed to various smokes and other obscurants during combat training. This report is intended to assist the Army in its efforts to ensure that exposures to these substances do not adversely effect the health of military personnel or the public living and working near military-training facilities. In this report, the National Research Council's Subcommittee on Military Smokes and Obscurants reviews the available toxicity data on four obscuring smokes—fog oil, diesel fuel, red phosphorus, and hexachloroethane—and develops exposure guidance levels for each.

The subcommittee was greatly assisted by several individuals who provided information on the uses and toxicity of the smokes considered in this report. We gratefully acknowledge Colonel Francis L. O'Donnell, Major James Martin, Lieutenant Colonel Forrest Oliverson, and the Office of the Surgeon General of the U.S. Army for their interest and support of this project. We also thank other persons who provided information for the subcommittee, including Winnifred Palmer, Sandra Thomson, Stephen Kistner, and Michael Burnham (all from the U.S. Army), Ian Greaves (University of Minnesota), David Gaylor (U.S. Food and Drug Administration), and Catherine Aranyi (IIT Research Institute).

We are grateful for the assistance of the NRC staff in the

preparation of this report. In particular, the subcommittee wishes to acknowledge Kulbir S. Bakshi, program director for the Committee on Toxicology, and Margaret E. McVey, project director for the subcommittee. Other staff members who contributed to this effort are Paul Gilman, executive director of the Commission on Life Sciences; James J. Reisa, director of the Board on Environmental Studies and Toxicology; Carol A. Maczka, program director for toxicology and risk assessment; Ruth E. Crossgrove, editor; and Lucy Fusco and Linda V. Leonard, project assistants.

Finally, we would like to thank all the members of the subcommittee for their expertise and dedicated effort throughout development of this report.

Michele A. Medinsky, Ph.D.

Chair, Subcommittee on Military Smokes and Obscurants

Rogene F. Henderson, Ph.D.

Chair, Committee on Toxicology

Contents

List of Abbreviations

ACGIH	American Conference of Governmental Industrial Hygienists
COT	Committee on Toxicology
CT	the product of concentration and time
DOD	U.S. Department of Defense
EEGL	emergency exposure guidance level
EPA	U.S. Environmental Protection Agency
FDA	U.S. Food and Drug Administration
HCE	hexachloroethane (the chemical)
HC	hexachloroethane (when combined with zinc oxide to produce a smoke)
HEPA	high-efficiency particulate air
IARC	International Agency for Research on Cancer
LC_{50}	lethal concentration for 50% of the test animals
LCT_{50}	lethal concentration multiplied by exposure time for 50% of the test animals
LD_{50}	lethal dose for 50% of the test animals
LOAEL	lowest-observed-adverse-effect level
MOUT	military operations in urban terrain
NIOSH	U.S. National Institute for Occupational Safety and Health
NOAEL	no-observed-adverse-effect level
NRC	National Research Council

OSHA	U.S. Occupational Safety and Health Administration
PAH	polycyclic aromatic hydrocarbons
PEL	permissible exposure limit
PEGL	permissible emergency guidance level
PPEGL	permissible public exposure guidance level
RP-BR	red phosphorus-butyl rubber
SPEGL	short-term public exposure guidance level
STEL	short-term exposure limit
TLV	Threshold Limit Value
TWA	time-weighted average
VEESS	vehicle-engine-exhaust-smoke system

Toxicity of Military Smokes and Obscurants

Summary

A variety of smokes and obscurants have been developed and used in wartime operations to screen armed forces from view, signal friendly forces, and identify enemy targets. Obscurants are anthropogenic or naturally occurring particles that are suspended in the air and block or weaken transmission of a particular part or parts of the electromagnetic spectrum, such as visible and infrared radiation or microwaves. Fog, mist, and dust are examples of obscurants. Smokes are produced by burning or vaporizing some product.

Large quantities of smokes and other obscurants are used in military training. The U.S. Army wishes to ensure that exposure to smokes and obscurants during training does not have adverse health effects on military personnel. To protect the health of exposed individuals, the Office of the Army Surgeon General requested that the National Research Council (NRC) review data on the toxicity of smokes and obscurants and recommend exposure guidance levels for military personnel in training and for the general public residing or working near military-training facilities.

The NRC assigned this project to the Committee on Toxicology (COT), which convened the Subcommittee on Military Smokes and Obscurants. The subcommittee conducted a detailed evaluation of data on the toxicity of four obscurant smokes: fog oil, diesel fuel, red phosphorus, and hexachloroethane. Toxicity data and exposure guidance levels for other smokes and obscurants will be presented in subsequent volumes.

The Army requested recommendations for four types of exposure limits: (1) emergency exposure guidance levels (EEGLs) for a rare, emergency situation resulting in exposure of military personnel for less than 24 hr; (2) permissible exposure guidance levels (PEGLs) for repeated exposure of military personnel during training exercises; (3) short-term public emergency guidance levels (SPEGLs) for a rare, emergency situation potentially resulting in an exposure of the public to military-training smoke; and (4) permissible public exposure guidance levels (PPEGLs) for repeated accidental exposures of the public residing or working near military-training facilities.

EXPOSURE GUIDANCE LEVELS FOR MILITARY PERSONNEL

Using NRC guidelines published in 1986 and 1992 for developing exposure guidance levels, the subcommittee developed EEGLs and PEGLs for the four obscuring smokes as shown in Table S-1 and described below.

Diesel-Fuel Smoke

Diesel-fuel smoke is formed by injecting diesel fuel into the exhaust manifold of a tactical vehicle. The fuel is vaporized and expelled with the vehicle's exhaust. The vapor condenses when exposed to the atmosphere, producing a visual obscurant composed of respirable particles.

Although extensive data are available on the health effects of combusted diesel-fuel exhaust, little information is available on the health effects of uncombusted diesel-fuel smoke. The mortality of rodents following one-time exposure depends on the product of exposure concentration and time (CT). One-time and repeated exposures to diesel-fuel smoke produce adverse effects in the respiratory tract of rats and mice. Toxic effects include pulmonary congestion, bronchopneumonia, bronchitis, edema, and hemorrhage. In several studies, diesel-fuel was neither neurotoxic nor genotoxic. In one developmental toxicity study, a slight delay in skeletal development was observed in rats; however, in several other studies, developmental or reproductive toxicity was not observed.

TABLE S-1 EEGLs and PEGLs for Smokes for Military Personnel

Smoke or Obscurant	Exposure Guideline	Exposure Duration	Guidance Level (mg/m^3)
Diesel-fuel smoke	EEGL	15 min	300
		1 hr	80
		6 hr	15
	PEGL	8 hr/d, 1 d/wk	10
		8 hr/d, 2 d/wk	5
Fog-oil smoke	EEGL	15 min	360
		1 hr	90
		6 hr	15
	PEGL	8 hr/d, 5 d/wk	5
Red phosphorus-butyl rubber smoke	EEGL	15 min	40
		1 hr	10
		6 hr	2
	PEGL	8 hr/d, 5 d/wk	1
Hexachloroethane smoke (as $ZnCl_2$)	EEGL	15 min	10
		1 hr	3
		6 hr	0.4
	PEGL	8 hr/d, 5 d/wk	0.2

Abbreviations: EEGL, emergency exposure guidance level; PEGL, permissible exposure guidance level.

For diesel-fuel smoke, the subcommittee developed EEGLs on the basis of an estimate of the CT product that induces a 1% mortality of rats following a single exposure for 2 to 6 hr (i.e., a CT product of 8,200 milligrams per cubic meter multiplied by hour (mg·hr/m^3)). Considering the severity of the end point (death), the subcommittee divided the CT product by an uncertainty factor of 10 to predict a nonpermanent health impairment and by another uncertainty factor of 10 to account for interspecies differences in sensitivity. The result is an EEGL (expressed as a CT product) of 80 mg·hr/m^3. Assuming that Haber's law applies in the absence of evidence to the contrary, the 15-min EEGL is 300 mg/m^3, the 1-hr EEGL is 80 mg/m^3, and the 6-hr EEGL is 15 mg/m^3.

The subcommittee based the PEGL on a lowest-observed-adverse-effect level (LOAEL) of 8,000 mg·hr/m^3 per week for focal pneumonitis in rats exposed for 9 weeks. The LOAEL was divided by an uncertainty factor of 10 to estimate a no-observed-adverse-effect level (NOAEL) and by another uncertainty factor of 10 to account for interspecies differences in sensitivity. The resulting PEGL, expressed as a CT product, is 80 mg·hr/m^3 per week. That PEGL corresponds to a PEGL of 10 mg/m^3 for one 8-hr exposure per week and 5 mg/m^3 for two 8-hr exposures per week. The subcommittee recommends those PEGLs as ceiling values; in other words, those PEGLs apply even if the exposure events are less than 8 hr in a given day. The subcommittee also recommends that protective equipment be worn if exposures to diesel-fuel smoke during training appears to produce chronic dermatitis in any individuals.

Fog-Oil Smoke

Fog-oil smoke is the term used for an oil smoke generated by injecting mineral oil into a heated manifold. As for diesel-fuel smoke, the fog-oil vapors condense when exposed to the atmosphere, producing respirable particles. The chemical and physical properties of fog oil are similar to those of petroleum-based lubricating and cutting oils.

In this report, the subcommittee distinguishes between "old" and "new" fog oil. Conventionally refined mineral oils, including old fog oil, can cause cancer of the skin of the arms, hands, and scrotum of humans. Polycyclic aromatic hydrocarbons (PAHs) and related compounds in conventionally refined mineral oils are thought to be responsible for those effects. In 1986, the military changed its specifications for fog oil and required severe solvent refining or severe hydro-treatment of fog oil to remove carcinogenic or potentially carcinogenic constituents. Severely refined fog oil is referred to as new fog oil.

Because of the carcinogenic properties of conventionally refined mineral oils, the subcommittee endorses existing Army recommendations that fog oil purchased before revision of the military specifications in 1986 no longer be used to produce smoke. The subcommittee also endorses Army recommendations that fog oil purchased after the specifications were revised be tested for carcinogenic constituents to ensure that all batches are free of carcinogens. The subcommittee's exposure guidance levels described below apply to new fog oils only.

The most sensitive toxic end point following short- and long-term exposures to new fog-oil aerosols in humans and animals appears to be respiratory-tract toxicity. To develop EEGLs, the subcommittee divided a 2-hr LOAEL of 4,500 mg/m^3 for pulmonary effects in mice by an uncertainty factor of 10 to estimate a NOAEL from a LOAEL and by another uncertainty factor of 10 to account for interspecies differences in sensitivity. The resultant EEGL is 45 mg/m^3 for 2 hr. Assuming Haber's law applies in the absence of evidence to the contrary, the 15-min EEGL is 360 mg/m^3, the 1-hr EEGL is 90 mg/m^3, and the 6-hr EEGL is 15 mg/m^3. The PEGL of 5 mg/m^3 is based on a study that indicated few, if any, complaints from workers exposed at or below that level.

Red Phosphorus Smoke

Red phosphorus smoke is deployed explosively from grenades and mortar shells. The obscurant portion of the grenades consists of a 95:5 mixture of red phosphorus and butyl rubber, which, when combusted, produces aerosols of phosphoric acid in a complex mixture of polymeric forms.

The high phosphoric acid content of the smoke causes respiratory-tract irritation and inflammation in humans and animals at concentrations of 180 mg/m^3. Inhalation of red phosphorus-butyl rubber smoke by rats produces terminal bronchiolar fibrosis. Induction of fibrosis appears to be influenced by both concentration and duration of exposure.

The most sensitive toxic effect following short-term exposures of humans and animals to red phosphorus-butyl rubber aerosols is respiratory distress. Concentrations as low as 100 mg/m^3 are considered to be intolerable to humans, even for short periods. Data from dogs and rats indicate that exposure to approximately 1,200 mg/m^3 for 1 hr induces respiratory distress. Dividing by an uncertainty factor of 10 to account for interspecies differences in sensitivity and by another uncertainty factor of 10 to estimate a NOAEL from a LOAEL, the subcommittee developed a 1-hr EEGL of 10 mg/m^3. Assuming Haber's law applies over relatively short exposure durations in the absence of evidence to the contrary, the 15-min EEGL is 40 mg/m^3, and the 6-hr EEGL is 2 mg/m^3.

The PEGL recommended for red phosphorus-butyl rubber is based on the American Conference of Governmental Industrial Hygienist's (ACGIH) Threshold Limit Value (TLV) time-weighted average (TWA) for phosphoric acid, which is the primary combustion product of concern. The TLV-TWA of 1.0 mg/m^3 appears to protect occupational workers adequately and, therefore, seems appropriate for military personnel as well.

Hexachloroethane Smoke

Hexachloroethane (HCE) smoke (often referred to as HC smoke) is produced by burning a mixture containing roughly equal parts of HCE and zinc oxide and approximately 6% granular aluminum.

The toxicity of HC smoke is attributed to the production of zinc chloride ($ZnCl_2$). Fatalities have occurred when military personnel were exposed to the discharge of HC smoke devices in enclosed spaces.

Inhalation of HC smoke causes respiratory effects in humans and animals. Data from humans indicate a threshold for slight nausea and irritation of the nose, throat, and chest from exposure to HC smoke with CT products of $ZnCl_2$ between 160 and 240 mg•min/m^3. With CT products at 1,700 mg•min/m^3 and above, effects can be severe and require hospitalization and treatment. Data from animals are sparse but indicate a NOAEL for HC smoke with $ZnCl_2$ at 26.6 mg/m^3 in rodents for daily 1-hr exposures and a LOAEL for HC smoke with $ZnCl_2$ at 254 mg/m^3 for inflammatory changes in the lung and death, suggesting a relatively steep dose-response curve.

HC smoke has been reported to produce alveolar carcinomas in mice. Fitting a generalized multistage linear dose-response model to those data provides an upper limit of the cancer risk of 0.086 per milligram $ZnCl_2$ per kilogram of body weight per day.

To establish EEGLs, the subcommittee used the CT product threshold of 160-mg•min/m^3 for nausea and respiratory irritation in humans as an acceptable exposure level for short-term emergencies. Applying Haber's law to the CT product of 160 mg•min/m^3, the 15-min EEGL is 10 mg/m^3, the 1-hr EEGL is 3 mg/m^3, and the 6-hr EEGL is 0.4 mg/m^3 expressed as milligrams of $ZnCl_2$.

Virtually no human data are available to estimate a PEGL for HC smoke. Dividing the rodent NOAEL of 26.6 mg/m^3 by an uncertainty factor of 10 to account for the shorter daily exposures of the test animals than of military personnel and by another uncertainty factor of 10 to account for interspecies differences in sensitivity, the subcommittee developed a PEGL of 0.2 mg/m^3 for 8 hr per day, 5 days per week.

Because one study suggested that HC smoke is carcinogenic in mice, the subcommittee derived a cancer potency factor for HC smoke to determine whether its potential carcinogenicity required further attention. The subcommittee found that the possible cancer risks associated with the recommended EEGLs and PEGLs were approximately 1 in a million. Moreover, using actual air-concentration data from a real training facility and using unrealistic worst-case weather conditions, the subcommittee estimated risks to the community closely surrounding that facility to be less than 1 in a million.

EXPOSURE GUIDANCE LEVELS AT BOUNDARIES OF MILITARY-TRAINING FACILITIES

The subcommittee developed SPEGLs and PPEGLs to ensure the protection of communities living near the facilities (Table S-2). In developing SPEGLs and PPEGLs, the subcommittee assumed that the general population includes sensitive subpopulations, such as the elderly, pregnant women, infants, children, and the chronically ill. In the absence of direct information on the toxicity of the smokes and obscurants in sensitive subpopulations, the subcommittee recommends that an uncertainty factor of 10 be used to extrapolate from guidance exposure levels derived for a population of healthy adults in the military to levels protective of more sensitive human subpopulations.

For all four obscurant smokes evaluated in this volume, the SPEGLs were estimated by dividing the EEGLs by an uncertainty factor of 10 to account for the likelihood of sensitive subpopulations in nearby communities. In addition, for all four smokes, the PPEGLs were estimated from the PEGLs and divided by an uncertainty factor of 10, again to account for the possibility that more sensitive subpopulations might reside near a military-training facility. Thus, all SPEGLs and PPEGLs in Table S-2 are 0.1 times the corresponding EEGLs and PEGLs in Table S-1.

TABLE S-2 SPEGLs and PPEGLs for Smokes at Boundaries of Military-Training Facilities

Smoke or Obscurant	Exposure Guideline	Exposure Duration	Guidance Level (mg/m^3)
Diesel-fuel smoke	SPEGL	15 min	30
		1 hr	8.0
		6 hr	1.5
	PPEGL	8 hr/d, 1 d/wk	1
		8 hr/d, 2 d/wk	0.5
Fog-oil smoke	SPEGL	15 min	36
		1 hr	9
		6 hr	1.5
	PPEGL	8 hr/d, 5 d/wk	0.5
Red phosphorusbutyl rubber smoke	SPEGL	15 min	4
		1 hr	1
		6 hr	0.2
	PPEGL	8 hr/d, 5 d/wk	0.1
Hexachloroethane smoke	SPEGL	15 min	1
		1 hr	0.3
		6 hr	0.04
	PPEGL	8 hr/d, 5 d/wk	0.02

Abbreviations: SPEGL, short-term public emergency guidance level; PPEGL, permissible public exposure guidance level.

1

Introduction

Ever since smokeless powder replaced black powder as the standard propellant for guns and firearms, armed forces have sought methods to create a haze similar to that created by black powder to blanket battlefields. A variety of smokes and obscurants have been developed and used in wartime operations for screening armed forces from view, deceiving the enemy, signaling friendly forces, and identifying enemy targets. Obscurants are used to hinder target acquisition, visual communication, and movement by the enemy. White to gray smokes deployed in grenades are used to cover or screen individual vehicles, and colored-smoke grenades are used to mark specific locations. Smokes can be deployed near the enemy to obscure or interfere with the enemy's vision or deployed in the area of friendly forces to screen or conceal the location and activities of the forces. The use of smokes and obscurants by the armed forces is essential for the achievement of tactical goals during wartime. To ensure defense preparedness, large quantities of smokes and obscurants also are used in military training areas.

Obscurants are anthropogenic or naturally occurring particles suspended in the air that block or weaken the transmission of a particular part or parts of the electromagnetic spectrum, such as visible and infrared radiation or microwaves. Fog, mist, and dust are examples of obscurants. Smoke is an obscurant normally produced by burning or vaporizing some product.

The selection of obscurants used during a military operation depends on the tactical needs. For example, visual obscurants, which block visible light and the near-infrared portion of the electromagnetic spectrum, are used to block viewers, such as binoculars, weapon sights, night-observation sights, and laser-range finders. Bispectrual obscurants, which also block light in the far-infrared portion of the spectrum, hinder battlefield viewers and weapons-guidance systems, such as homing systems on antitank- and air-defense missiles. Other multispectrual obscurants can defeat radar systems and high-energy microwave-directed weapons.

SMOKES REVIEWED IN THIS REPORT

The effectiveness of obscuring and screening smokes depends on their ability to obscure visibility by reflecting, refracting, and scattering light rays. For this reason, all such military smokes consist of aerosols with particle dimensions approximating the wavelength of visible to near-infrared light. The relationship between smoke concentration and the associated reduction in visibility is summarized in Table 1-1 for selected smokes. As the smoke concentration increases, the obscurant effectiveness increases as indicated by the decreasing visibility.

Among the most widely used smoke munitions in actual combat are those that produce smoke by burning a mixture of chlorinated hydrocarbons and zinc oxide (ZnO). The hydrocarbon generally used in these smoke mixtures is hexachloroethane (HCE), and the generic name the military attaches to the HCE-ZnO smokes is hexachloroethane smoke or HC smoke. Most of the HCE-ZnO mixture produced for the military is used in smoke pots or cylindrical metal canisters containing the HCE-ZnO mixture along with a pyrotechnic charge that, when ignited, provides the heat necessary to generate the HC smoke. Smoke pots ordinarily are used to provide small-area screens, supplement other smoke sources by filling holes in screens, and help establish screens rapidly.

TABLE 1-1 Correlation Between Visibility and Smoke Concentration for Selected Smokes

Smoke	Visibility[a] (m)	Concentration (mg/m^3)
Fog-oil (SGF-2) smoke	10	31
	50	6.2
	200	1.6
Diesel-fuel (DF-2) smoke	10	39
	50	7.9
	200	2.0
Red phosphorus smoke	10	62
(50% relative humidity)	50	12
	200	3.1
Hexachloroethane smoke	10	69
(85% relative humidity)	50	14
	200	3.5

[a] Visibility is defined as the path length for a 10% transmission at the concentration determined by the Beer-Lambert law.
Abbreviations: m, meter; mg/m^3, milligram per cubic meter.
Source: Adapted from Eaton and Young (1989).

Red phosphorus is incorporated into certain smoke grenades used for self-protection on armored vehicles. This munition ignites while airborne and obscures the vehicle from which it is launched. During grenade production, red phosphorus is plasticized with butyl rubber in an organic solvent and extruded. The extruded granules, which consist of a 95:5 mixture of red phosphorus and butyl rubber (RP-BR), are dried, pressed into pellets, and then inserted into the rubber sleeve of a grenade. Phosphorus artillery and mortar smoke rounds are used primarily for projecting smoke on the enemy, and the smoke screen can be greater than 500 m long. These smoke rounds are usually fired onto the munition impact zones. Thus, troop smoke exposure is probably minimal because troops generally do not train near munition impact zones.

One petroleum product used by the military to create an obscurant smoke is fog oil (a mineral oil). The smoke is generated by injecting fog oil into a heated manifold where it vaporizes as it is released into the atmosphere. The vapor quickly condenses on cooling in the air stream, creating an obscuring aerosol or mist. Oil mists created in that way are composed predominantly of respirable droplets. Because fog-oil smoke is used as a screen to reduce enemy observation of troop activity under the screen, exposure is unavoidable in areas where it is used. During World War II, for example, troops were exposed continuously to fog-oil smoke for up to 45 days. Today, certain Army personnel (the training instructors) might be exposed to fog-oil smoke for as long as 6 hr per week for an average of 6 years.

Another petroleum-based smoke used in military operations is diesel-fuel smoke. This smoke is produced by vaporizing diesel fuel within a vehicle-engine-exhaust-smoke system (VEESS). The resulting condensation smoke is used to screen armored vehicles from enemy view up to a distance of 1-2 kilometers (km). The VEESS is mounted on most armored vehicles and can produce smoke rapidly. However, VEESS-produced diesel-fuel smokes can pose a danger to troops on the ground and hinder the location of enemy targets.

U.S. ARMY POLICY CONCERNING USE OF OBSCURANTS

Current U.S. Army policy for training military personnel involving smokes or other obscurants requires that personnel carry protective masks when participating in exercises that include the use of obscurants (Eckelbarger 1985). Personnel are required to wear protective masks[1] in the following situations:

[1] Protective masks are full-faced air-purifying respirators that use charcoal and a HEPA filter (a separate air supply is not included).

1. Before exposure to any concentration of smoke produced by M8 white-smoke grenades, smoke pots (HC smoke), or metallic powder obscurants.
2. When passing through or operating in dense smoke (visibility less than 50 m).
3. When operating in or passing through a smoke haze (visibility greater than 50 m) that will exceed 4 hr in duration.
4. When exposure to smoke produces breathing difficulty, eye irritation, or discomfort. (Such effects in one individual serve as a signal for all similarly exposed personnel to wear masks.)
5. When using smoke during Military Operations in Urban Terrain (MOUT) training in enclosed spaces. (The Army policy notes that the protective mask is not effective in oxygen-deficient atmospheres; personnel are advised not to enter confined spaces where oxygen might have been displaced.)
6. In addition, smoke-generator personnel must wear masks if they cannot stay upwind of the smoke generator.

The U.S. Army policy also requires showering and laundering of clothing following training exercises to eliminate the risk of skin irritation from exposure to smoke. Troops exposed to smoke are advised to reduce skin exposure by rolling down sleeves. The policy requires that special care be taken when using HC smoke in training to ensure that appropriate protection is provided to all personnel who are likely to be exposed. Specific consideration must be given to weather conditions and the potential downwind effects of the smoke used in training. Positive controls, such as observation, control points, and communications, must be established to protect personnel who are not wearing masks from exposure to the smoke.

SUBCOMMITTEE TASK

The U.S. Army Medical Department wishes to ensure that exposure to smokes and other obscurants during combat training will not have adverse health effects on military personnel. The primary routes of exposure for soldiers are inhalation and dermal contact. The Office of the Army Surgeon General requested that the National Research Council (NRC) review the data on the toxicity of military smokes and obscurants and recommend exposure guidance levels for military personnel during combat training and for the general public residing or working near military-training facilities.

The NRC assigned this project to the Committee on Toxicology (COT), which convened the Subcommittee on Military Smokes and Obscurants. For this report, the subcommittee evaluated four obscuring smokes: fog oil, diesel fuel, red phosphorus, and hexachloroethane. Exposure guidance levels levels for other smokes and obscurants will be presented in subsequent volumes.

The task of the subcommittee was to review the health effects associated with exposure to the smokes and other obscurants and to recommend four exposure guidance levels: (1) emergency exposure guidance levels (EEGLs) for a rare, emergency situation resulting in an exposure of military personnel for less than 24 hr; (2) permissible exposure guidance levels (PEGLs) for repeated exposure of military personnel during training; (3) short-term public emergency guidance levels (SPEGLs) for a rare, emergency situation potentially resulting in an exposure of the public to a military-training smoke; and (4) permissible public exposure guidance levels (PPEGLs) for possible repeated exposures of the public residing or working near military-training facilities. All four guidance levels should take into account embryo and fetal development and reproductive toxicity in men and women. In addition, exposures of potentially sensitive subpopulations (e.g., ill or elderly persons and children) are considered in the SPEGL and PPEGL.

DEFINITIONS OF EXPOSURE GUIDANCE LEVELS

An EEGL is defined as a concentration of a substance in air (as a gas, vapor, or aerosol) that will permit continued performance of specific tasks during emergency exposures lasting up to 24 hr—an occurrence expected to be infrequent in the lifetime of a person (NRC 1986, 1992a). "Emergency" connotes a rare and unexpected situation with potential for significant loss of life, property, or mission accomplishment if not controlled. An EEGL, a single ceiling-exposure concentration for a specified duration, specifies and reflects the subcommittee's interpretation of available information in the context of an emergency.

An EEGL is acceptable only in an emergency, when some risks or some discomfort must be endured to prevent greater risks (such as fire, explosion, or massive release). Exposure at the EEGL might produce such effects as increased respiratory rate, headache, mild central-nervous-system effects, and respiratory-tract or eye irritation. The EEGL should prevent irreversible harm. Even though some reduction in performance is permissible, it should not prevent proper responses to the emergency (such as shutting off a valve, closing a hatch, or using a fire extinguisher). For example, in normal work situations, a degree of upper-respiratory-tract irritation or eye irritation causing discomfort would not be considered acceptable; during an emergency, it would be acceptable if it did not cause irreversible harm or seriously affect judgment or performance. The EEGL for a substance represents the subcommittee's judgment based on evaluation of experimental and epidemiological data, mechanisms of injury, and, when possible, operating conditions in which an emergency exposure might occur, as well as consideration of U.S. Department of Defense (DOD) goals and objectives. EEGLs were developed for military use and are intended for healthy, young military personnel. Therefore, they are not directly applicable to general populations consisting of elderly, very young, and ill persons.

A SPEGL is defined as a concentration of a substance in air that is acceptable for an unpredicted, single or rare, short-term emergency exposure of the general public. The SPEGL takes into account the likely wide range of susceptibility among individuals in the general public, including potentially sensitive subgroups, such as children, the elderly, and persons with serious debilitating diseases. Effects of exposure on the developing embryo and fetus and on the reproductive capacity of men and women also are considered in setting a SPEGL.

For purposes of assessing military smokes and other obscurants for the Army, the subcommittee developed two additional guidance levels, PEGLs and PPEGLs. The subcommittee defines a PEGL as the concentration of a substance in air to which healthy military personnel can be exposed repeatedly, up to a specified total exposure on a weekly basis (usually 8 hr per day, 5 days per week), for several years without experiencing adverse health effects or degradation in performance. The PEGL is similar to guidelines for occupational exposures, although the duration of exposure specified can be more or less than 40 hr per week, depending on military training regimens.

The subcommittee defines a PPEGL as the concentration of a substance in air to which the general public can be exposed repeatedly without experiencing any adverse health effects or discomfort. PPEGLs, like SPEGLs, take into account the likely wide range of susceptibility among individuals in the general public, including potentially sensitive subgroups (children, the elderly, and the chronically ill) and the developing embryo and fetus.

Exposure guidance levels developed by the subcommittee can be compared with other potentially useful exposure limits, such as the Threshold Limit Value (TLV) time-weighted averages (TWAs) and short-term exposure limits (STELs) recommended by the American Conference of Governmental Industrial Hygienists (ACGIH) for permissible workplace exposures (ACGIH 1991, 1995). These guidelines are developed for daily occupational exposures of healthy workers. The Occupational Safety and Health

Administration (OSHA) is responsible for promulgating and enforcing health standards in the majority of the work environments. These legally binding standards are referred to as permissible exposure limits (PELs; DOL 1995), and most follow the National Institute of Occupational Safety and Health (NIOSH) recommended exposure levels (RELs; EPA 1987). The OSHA and ACGIH values are not relevant for the general public.

Definitions of Exposure Guidance Levels for Military Smokes and Obscurants

EEGL	Emergency exposure guidance level for a rare, emergency situation resulting in an unanticipated exposure of military personnel for less than 24 hr.
PEGL	Permissible exposure guidance level for repeated exposure of military personnel during training.
SPEGL	Short-term public emergency guidance level for a rare, emergency situation potentially resulting in an exposure of the general public to military-training smoke.
PPEGL	Permissible public exposure guidance level for possible repeated exposures of the general public residing or working near military-training facilities.

APPROACH TO DEVELOPING EXPOSURE GUIDANCE LEVELS

The NRC has published guidelines for developing EEGLs, SPEGLs, and other exposure guidance levels for continuous or repeated exposures (NRC 1992a,b, 1996). For purposes of assessing military smokes and other obscurants, the subcommittee developed comparable procedures for developing PEGLs and PPEGLs.

The steps in developing exposure guidance levels are similar for EEGLs, SPEGLs, PEGLs, and PPEGLs; the differences reflect attributes of the exposed populations and the duration and frequency of exposure. The remainder of this section reviews the steps for developing an EEGL (NRC 1986) and then explains how procedures differ for the remaining three types of exposure guidance levels.

Emergency Exposure Guidance Levels (EEGLs)

The first step in developing an EEGL is to review all available toxicology information and any documentation for exposure limits proposed by the ACGIH and regulatory agencies. Acute toxicity is the primary basis for establishing an EEGL. If there is any evidence that the substance under consideration is carcinogenic in either animals or humans, a cancer risk assessment is performed to estimate the possible potency of the substance as a carcinogen. The approach used to estimate potency is developed case by case, depending on available data and plausible mechanisms of action.

In estimating an EEGL for a substance that has multiple biological effects, all end points are evaluated, and the most important is selected. In general, EEGLs reflect experimental and clinical observations and epidemiological, physiological, and toxicological data on animals and humans; both immediate and delayed health effects are considered. Special attention is given to training and battlefield conditions of concern to the military.

When developing an EEGL for a noncarcinogenic substance, the subcommittee first assesses the available and relevant toxicological information to determine the no-observed-adverse-effect level (NOAEL) of each smoke for the most sensitive end point. The NOAEL is the highest concentration for which there are data indicating that the smoke produced no adverse toxic effect. If a NOAEL cannot be determined from the available data, the lowest-observed-adverse-effect level (LOAEL) is determined. The LOAEL is the lowest level at which an adverse effect is seen in either human or laboratory animal studies. To estimate a NOAEL from a LOAEL, the subcommittee generally divides the LOAEL by a default uncertainty factor of 10, following the recommendations of the U.S. Environmental Protection Agency (EPA 1990). When NOAEL values obtained from laboratory animals are used to estimate exposure guidelines for humans, the subcommittee adopts the NRC Safe Drinking Water Committee (NRC 1977) default assumption that humans are 10-fold more sensitive than animals unless data are available that justify using a different assumption.

The development of an EEGL for different durations of exposure usually begins with the shortest exposure anticipated and works up to the longest. If there are no data for a given exposure duration, the subcommittee evaluates whether Haber's law can be applied to estimate the EEGL for that duration from an EEGL for a different duration. Haber's law states that a toxic-effect level is directly proportional to exposure concentration (C) multiplied by exposure time (T) over relatively short periods. In other words, for a constant product of C and T, the same effect level (e.g., 50% mortality) would result.

For many substances, the CT concept is inappropriate; therefore, each substance is evaluated for the likely applicability of Haber's law. The subcommittee employs Haber's law when data are available indicating a consistent relationship between relevant effects and the product of C and T. In some circumstances, the subcommittee applies Haber's law in the absence of clear supporting data. However, the use of Haber's law in the absence of supporting data is limited to (1) extrapolations between short-term exposures (e.g., 1 hr to 15 min; 4 hr to 1 hr) and (2) judgments of the subcommittee that the exposure-response relationship would follow Haber's law because of the physical and chemical properties of the material, the nature of the effects seen, and data on similar materials.

The subcommittee prefers to develop an EEGL for the short-term exposures first and then to use Haber's law to extrapolate to the longer-term EEGLs. Haber's law is used with caution to extrapolate from longer-term EEGLs to shorter-term EEGLs. The subcommittee develops EEGLs for three exposure durations—15 min, 1 hr, and 6 hr—to provide guidance for a range of potential emergency exposure conditions in the military-training environment.

Short-Term Public Emergency Guidance Levels (SPEGLs)

SPEGLs are generally set at 0.1 to 0.5 times the EEGL to protect sensitive subgroups, including infants, the elderly, and the chronically ill (NRC 1986).

Permissible Exposure Guidance Levels (PEGLs) and Permissible Public Exposure Guidance Levels (PPEGLs)

Although the NRC has not published guidelines for developing PEGLs and PPEGLs, the subcommittee follows the same approach as that recommended for EEGLs and SPEGLs, with modifications that reflect the repeated nature of exposures. In contrast to EEGLs and SPEGLs, the subcommittee uses chronic toxicity as the primary basis for establishing PEGLs and PPEGLs. If there is any evidence that the substance is carcinogenic in either animals or humans, the subcommittee estimates the possible potency of the compound as a carcinogen. DOD can use the potency value when comparing risks associated with different levels of exposure with risks incurred by personnel wearing masks or not using the obscurant. The subcommittee generally sets PPEGLs at 0.1 times the SPEGLs to protect more sensitive subgroups in the general public.

SUMMARY OF APPROACH

The general approach for developing EEGLs for a noncarcinogenic smoke is to determine a NOAEL directly from short-exposure laboratory experiments or human studies or to estimate an acute NOAEL from a LOAEL using a default assumption. A 15-min, 1-hr, or 6-hr EEGL is estimated from the NOAEL, depending on which exposure duration most closely matched the exposure duration associated with the NOAEL. Haber's law then is used, if appropriate, to develop EEGLs for different exposure durations.

SPEGLs generally are developed by taking the EEGL for healthy military personnel and applying an additional uncertainty factor to protect all members of the public, including sensitive subgroups, such as the elderly, children, and the developing embryo or fetus. In the absence of specific data on variation in human sensitivity to a smoke, the subcommittee assumes that some subgroups could be up to 10 times more sensitive than healthy military personnel. Thus, unless otherwise noted in this report, the SPEGLs are generally 10-fold lower than the EEGLs.

Data from chronic or repeated-exposure experiments or clinical observations form the basis for the PEGL instead of the acute-exposure data used for the EEGL. Haber's law is not applicable to the longer exposure durations.

The PPEGL for possible repeated exposures of a community near a military-training facility is developed by dividing the PEGL established for military personnel by an uncertainty factor of 10 to extrapolate from healthy military personnel to a more diverse population, including sensitive subgroups.

It is the aerosol (particulate) nature of the smokes covered in this volume that is responsible for their visual obscurant properties. Although some portion of some smokes might be present as a vapor rather than as an aerosol, the mass of the smoke in the air is overwhelmingly in the form of an aerosol if the smoke-generating equipment is operating correctly. All recommended exposure guidance values are reported in milligrams of total particulates per cubic meter. To the extent that vapors in equilibrium with the aerosol might contribute to the toxicity of the smoke, the measurement of milligrams of total particulates per cubic meter serves as a surrogate measurement for the combined aerosol and vapor components in the test situations and in the field.

The remainder of this report is organized in four chapters, one for each of the obscuring smokes evaluated in this volume. For each, the chapter presents background information on military applications and physical and chemical properties of the smoke. Each chapter also includes a discussion of toxicokinetics and a summary of the available toxicity data on the smoke. Following a description of existing recommended exposure limits, each chapter presents the subcommittee's evaluation of the toxicity data and development of the exposure guidance levels. Sections on research needs and references conclude each chapter.

REFERENCES

ACGIH (American Conference of Governmental Industrial Hygienists). 1991. Documentation of the Threshold Limit Values and Biological Exposure Indices, 6th Ed. American Conference of Governmental Industrial Hygienists, Cincinnati, Ohio.

ACGIH (American Conference of Governmental Industrial Hygienists). 1995. 1995-1996 Threshold Limit Values and Biological Exposure Indices. American Conference of Governmental Industrial Hygienists, Cincinnati, Ohio.

DOL (U.S. Department of Labor). 1995. Air Contaminants—Permissible Exposure Limits. Title 29, Code of Federal Regulations, Part 1910, Section 1910.1000. Washington, D.C.: U.S. Government Printing Office.

Eaton, J.C., and J.Y. Young. 1989. P. 11 in Medical Criteria for Respiratory Protection in Smoke: The Effectiveness of the Military Protective Mask. Tech. Rep. 8902. U.S. Army Biomedical Research and Development Laboratory, Fort Detrick, Frederick, Md.

Eckelbarger, M.G. 1985. Smoke Safety. Message from the Director of Army Safety, U.S. Department of the Army, Washington, D.C.

EPA (U.S. Environmental Protection Agency). 1987. Technical Guidance for Hazards Analysis: Emergency Planning for Extremely Hazardous Substances. Prepared by the U.S. Environmental Protection

Agency in conjunction with the Federal Emergency Management Agency and the U.S. Department of Transportation, Washington, D.C. Available from NTIS, Springfield, Va., Doc. No. PB93-206910.

EPA (U.S. Environmental Protection Agency). 1990. Interim Methods for Development of Inhalation Reference Concentrations. EPA 600/8-90/066A. Environmental Criteria and Assessment Office, U.S. Environmental Protection Agency, Research Triangle Park, N.C.

NRC (National Research Council). 1977. Drinking Water and Health. Vol. 1. Washington, D.C.: National Academy Press.

NRC (National Research Council). 1986. Criteria and Methods for Preparing Emergency Exposure Guidance Level (EEGL), Short-term Public Emergency Guidance Level (SPEGL), and Continuous Exposure Guidance Level (CEGL) Documents. Washington, D.C.: National Academy Press.

NRC (National Research Council). 1992a. Guidelines for Developing Community Emergency Exposure Levels for Hazardous Substances. Washington, D.C.: National Academy Press.

NRC (National Research Council). 1992b. Guidelines for Developing Spacecraft Maximum Allowable Concentrations for Space Station Contaminants. Washington, D.C.: National Academy Press.

NRC (National Research Council). 1996. Toxicity of Alternatives to Chlorofluorocarbons: HFC-134a and HCFC-123. Washington, D.C.: National Academy Press.

2

Diesel-Fuel Smoke

BACKGROUND INFORMATION

Military Applications

Diesel-fuel smoke is one of the visual obscurants used by the armed forces to conceal personnel and equipment. Diesel-fuel smoke is formed by injecting diesel fuel into the exhaust manifold of a tactical vehicle where the fuel is vaporized and expelled with the vehicle's exhaust. Upon dilution and cooling to the ambient temperature, the fuel condenses into a dense white smoke. Because military personnel might be exposed to this aerosol in routine training situations and in actual combat, its effects on their performance and health are of interest.

Physical and Chemical Properties

Diesel fuel is a mixture of aliphatic, olefinic, and aromatic hydrocarbons obtained from the distillation of petroleum. The composition of the fuel purchased by the U.S. Army to generate smoke is controlled only by specifications on boiling point, cetane number (a measurement of ignition quality), viscosity, and flash point. Additives might also be present in small quantities to improve combustibility (alkyl nitrates), reduce corrosion of storage vessels (surfactant), or reduce gum formation (antioxidants such as aromatic amides or phenols).

Diesel fuels are categorized as the middle distillates from crude oil and are more dense than gasoline. As defined in the U.S. Chemical Substances Inventory under the Toxic Substances Control Act, diesel fuels consist of hydrocarbons with carbon numbers predominantly in the range of C_9 to C_{20} and boils in the range of 163 to 357°C (IARC 1989). That definition encompasses both diesel fuels No. 1 (DF1) and No. 2 (DF2). DF1 is essentially kerosene and consists of hydrocarbons with numbers predominantly in the range of C_9 to C_{16} and boiling in the range of 150 to 300°C. DF1 contains little benzene (e.g., less than 0.02%) or polycyclic aromatic hydrocarbons (Millner et al. 1991). DF2 is essentially equivalent to fuel oil No. 2 used for automobiles and boils between 160 and 360°C (IARC 1989). DF2 is more viscous than DF1 and spans a carbon number range of C_{11} to C_{20} (IARC 1989). DF2 also contains a greater variety of compounds and includes olefins and mixed aromatic olefin-type compounds, such as styrenes. More information on the composition of diesel fuels can be found in IARC (1989) and Millner et al. (1992).

The characteristics of DF2 (Jenkins et al. 1983a) are the following:

Composition		
	Saturated hydrocarbons	70.0%
	Substituted benzenes	16.0%
	2-Ring aromatics	12.0%
	3-Ring aromatics	2.0%
	Polar aromatics	0.2%
Refractive index		1.477
Density at 25°C (grams per milliliter)		0.844
Viscosity at 25°C (centistokes)		3.35

Flash point (°C minimum) Distillation range		74.0
	10% Point, °C	186.0
	50% Point, °C	271.0
	90% Point, °C	298.0

Diesel-fuel smoke—technically, a fog—is a condensation aerosol, consisting of a suspension of 0.5 to 1.0 micrometer (μm) fuel droplets in air. The droplets are individually translucent but opaque en mass. Particles in this range are respirable. The generation of the condensation aerosol is for the purpose of obscuring the soldiers from view; if conditions were such that a major portion of the fuel remained in the vapor form, the system would not achieve its purpose. However, a fraction of the fuel (components with low boiling points) might remain in the vapor form. Combusted fuel might also contribute to the total mass of the smoke, but Callahan et al. (1983) found that vehicle exhaust contributed only 1% to 2% of the total hydrocarbon concentration of the smoke. All measurements of and recommendations for diesel-fuel-smoke concentrations are reported as milligrams of total particulates per cubic meter.

TOXICOKINETICS

Although no toxicokinetic studies have been conducted with diesel-fuel smoke, some evidence of the deposition and clearance pattern of the smoke is available from studies of Dalbey et al. (1982, 1987) and Lock et al. (1984). In these studies (described in detail in the Toxicity Summary below), the lung was the primary target organ, with several indicators of an inflammatory response. The material appears to deposit and accumulate in the lung, remaining there long enough to induce an inflammatory response. Within a 2-week period, neutrophil levels are back to control levels, suggesting that most of the particles are cleared. However, the macrophage levels are still elevated after that time period, indicating that some inhaled particles might still be in the lung. Dalbey et al. (1982, 1987) and Dalbey and Lock (1982) made use of dodecachlorobiphenyl as a dosimetric tracer for aerosols of diesel-fuel (Jenkins et al. 1983b). The fraction of inhaled diesel-fuel aerosol retained by the rats at the end of exposure was 4% to 8%. The largest internal amounts of the tracer were found in the lungs. Animals exposed to CT products at 8,000 milligrams per cubic meter multiplied by hour (mg•hr/m^3) had between 2 and 4 mg of the aerosol particles in their lungs, and animals exposed to CT products at 12,000 mg•hr/m^3 had between 3 and 6 mg of particles in their lungs. Tracer found in the upper respiratory tract accounted for less than 1.5% of the total internal dose, and that found in the digestive system accounted for approximately 30% of the total internal dose.

TOXICITY SUMMARY

Effects in Humans

Although extensive data are available on the health effects of combusted diesel-fuel exhaust, little information is available on the health effects of uncombusted diesel-fuel smoke in humans. Volunteers who were exposed to concentrations of 170 and 330 mg/m^3 for 10 min reported no irritant effects (Dautrebande and Capps 1950). No other studies involving inhalation exposure of humans have been reported.

Repeated exposure to diesel fuel was reported to cause contact dermatitis in sensitive individuals (Il'in et al. 1969), but inadequate personal hygiene was indicated as a contributing factor. The literature regarding gastritis from ingested diesel fuel and pneumonia from aspiration of diesel fuel has been reviewed (Liss-Suter et al. 1978).

Effects in Animals

Dermal and Ocular Exposures

Lethality

Repeated application of commercial diesel fuel to the shaved skin of rabbits at 4 and 8 milliliters per kilogram (mL/kg) of body weight produced 0 and 67% mortality, respectively (Beck et al. 1982). Toxic signs included weight loss, anorexia, and severe dermal irritation. Necropsy revealed congested kidneys and livers.

Eye and Skin Irritation

Commercial diesel fuel did not induce eye irritation in rabbits during a 30-sec application or skin sensitization in guinea pigs (Beck et al. 1982). The fuel was, however, extremely irritating to the skin when applied for 24 hr.

Carcinogenic Effects

DF2 was positive when tested as a tumor promoter in the SENCAR mouse-skin tumorigenesis bioassay but was negative as a complete carcinogen in the same assay (Slaga et al. 1983).

Although no chronic in vivo bioassays have been conducted with diesel fuel per se, mouse skin-painting studies have found some middle distillate fractions of crude oil to be tumorigenic when applied to the skin (clipped free of fur) twice weekly for a lifetime (Blackburn et al. 1984, Lewis et al. 1984). However, Ingram and Grasso (1991) noted that the distillate fractions acted as irritants and, under the conditions of the chronic applications, caused overt skin damage, giving rise to epidermal hyperplasia, which might have enhanced the development of the tumors.

Inhalation Exposures

Two major sets of animal studies on the toxicity of inhaled diesel-fuel smokes have been produced; one was conducted at Aberdeen Proving Ground (Callahan et al. 1983, 1986) and one at Oak Ridge National Laboratory (Dalbey and Lock 1982; Dalbey et al. 1982, 1987; Lock et al. 1984). These studies cover both one-time and repeated exposures for up to 13 weeks. For one-time exposures, the product of exposure concentration and time is provided.

One-Time Exposures

Lethality. In a single-exposure study (Callahan et al. 1983), rats (Sprague-Dawley and Fischer 344), mice ($B6C3F_1$), and guinea pigs (Hartley) were exposed to high concentrations of DF1 or DF2 smoke and exhaust for 15, 60, 120, or 300 min under static airflow conditions. The goal was to expose the animals to 10,000 to 12,000 mg/m^3 of smoke, as the theoretical average smoke concentration predicted for exposure of military personnel standing within 10 m downwind from the tank, operating at maximal smoke-generating efficiency under strong inversion and low-wind-speed atmospheric conditions.

Generation and dissemination of the diesel-fuel smokes were accomplished by accelerating the 750-hp Chrysler engine of the tanks up to 1,700 revolutions per minute (rpm) for approximately 5 min. When the engine manifold temperature reached 1,180°F, the fuel was expelled into the manifold through an orifice and the generated smoke was drawn through a stainless steel tube into a wind tunnel. After a generation period of 5 min, the smoke was shunted into a 20,000-liter cylindrical exposure chamber. Caged animals were placed in the chamber before the introduction of the smoke. For exposures exceeding 60 min, the generation procedure was repeated every 60 min because the hydrocarbon concentrations in the air of the exposure chamber decreased by 65% in 60 min.

The average concentration of the DF2 smoke for the 15- and 60-min exposures was 35,000 mg/m^3 . The average concentration for the DF1 smoke was 15,000 mg/m^3 for exposures from 15 to 300 min. Because dissemination of smoke generated from the M60A1 tank also involves normal exhaust emission from internal engine combustion, the concentration of the exhaust component was monitored. The hydrocarbon concentration from the exhaust of the engine was only 1% to 2% of the total smoke plus exhaust hydrocarbon concentration. The mass median aerodynamic diameter of the smoke particles was 0.23 µm or less (measured by cascade impaction at concentrations of 15,000 to 18,000 mg/m^3).

Gases were also monitored in the exposure chambers and were reported as not exceeding the following values: NO_2, 40 parts per million (ppm); SO_2, 20 ppm; CO, 130 ppm; and CO_2, 11,000 ppm. Oxygen never dropped below 18.9%.

Animals could not be observed during the exposures because of the density of the smoke. At the end of the 15-min exposures to DF2 smoke and exhaust, all animals but the mice appeared lethargic. The 1-hr exposures resulted in lacrimation, oral and nasal frothiness, nasal hemorrhage, and tremors in the guinea pigs. Sprague-Dawley rats manifested cyanosis, and mice showed piloerection. One of 10 F344 rats and 4 of 10 guinea pigs died from the 60-min exposure. Mortality of animals exposed to DF1 smoke increased with increasing length of exposure; after 300 min of exposure, all but 4 mice (out of 10 mice per strain) died.

Dalbey and Lock (1982) conducted experiments to establish the maximum tolerated concentration for a given exposure duration in male and female Sprague-Dawley rats. The generation of the aerosol, which was designed to model the vehicle exhaust system used by the military to produce smoke from diesel fuel, consisted of a 1-in. O.D. stainless steel tube about 1 m long, with a Vycor heater fitted into one end. The heater was maintained at 600°C. The distal end of the generator was heated to 350°C by heating tape. Nitrogen entered the tube near the Vycor heater and exited at the opposite end. Diesel fuel was metered onto the tip of the Vycor heater, where it was flash evaporated and carried by the hot nitrogen out of the generator into the cool supply air entering the exposure chamber. Aerosol concentration in the chamber was monitored continuously by infrared backscatter probes at the top and bottom of the chamber. Particle size was determined by cascade impaction at random intervals during the study. The mass median aerodynamic diameter of the particles ranged from 0.43 to 0.75 μm with a geometric standard deviation of 1.4 to 1.5. Air flow through the chamber was maintained at 425 L/m, and aerosol distribution within the chamber was uniform.

Exposures were for 2, 4, or 6 hr at concentrations ranging from 670 to 16,000 mg/m^3. Exposure groups consisted of five males and five females. Mortality was observed for 2 weeks after the exposures. Mortality was the same in both sexes, and the data were combined. Mortality was found to be highly correlated with C·T, the CT product explaining 83% of the variation in mortality. A probit analysis was used to relate mortality to the log of the CT product so that the CT product that induces 1% mortality could be estimated. That estimate was 12,200 $mg{\cdot}hr/m^3$, and the 97.5% lower confidence bound of that value was calculated to be 8,200 $mg{\cdot}hr/m^3$.

Pulmonary Effects. The histopathology studies conducted by Callahan et al. (1983, 1986) indicated that the major lesions in the exposed animals were in the respiratory tract. Lesions were found in the nasal turbinates and lungs, and included congestion in the nasal turbinate, bronchopneumonia, bronchitis, peribronchiolitis, peribronchiolar lymphocytosis, pulmonary histiocytosis, and pulmonary congestion with edema and hemorrhage.

Dalbey and Lock (1982) found upon necropsy of the rats in their experiments that the only significant gross abnormality was edema in the lungs and occasionally in the trachea.

Repeated 15- or 60-Minute Exposures

Callahan et al. (1986) exposed B6C3F_1 mice and F344 rats to an average of 2,300 mg/m^3 of M60A1 tank-generated exhaust and DF2 smoke for 15 or 60 min daily (5 days per week) for up to 13 weeks. Analysis of gas concentrations indicated an average of 4.4 ppm for NO_2, 7.5 ppm for SO_2, and 16 ppm for CO. CO_2 did not exceed 1,000 ppm, and O_2 did not go below 20%. The exposures were performed under static airflow conditions as described for the acute toxicity studies (Callahan et al. 1983). Toxicological, physiological, hematological, blood chemical, behavioral, reproductive, mutagenic, teratogenic, and pathological effects were evaluated.

The only clinical sign of toxicity observed in the exposed rodents was hypoactivity following the 32 or more consecutive daily exposures. After the daily exposure ended, hypoactivity diminished within 24 hr. Carboxyhemoglobin levels in the exposed rodents did not exceed 11%.

No gross pathological change was observed in exposed animals, but the incidence of mild-to-moderate pulmonary congestion increased as compared with controls. Exposed rats also had increased incidences of inflammatory and vascular lesions in nasal turbinates and tracheas. The respiratory-tract changes were of low severity and were not related to the duration of exposure. Therefore, the investigators did not consider the findings likely to be exposure-related.

Repeated 4-Hour Exposures

Lock et al. (1984) conducted a 13-week study to determine if there were cumulative toxic effects from repeated exposures to low concentrations of diesel-fuel smoke. Male and female Sprague-Dawley rats were exposed for 4 hr, twice a week for 13 weeks (26 exposures). The exposure pattern was meant to mimic what humans might experience in the field. Animals were observed either

5 days after the last exposure or after a recovery period of 2 months. Actual exposure concentrations of the smoke were 170, 870, and 1,600 mg/m^3, as measured by infrared backscatter probes and averaged over the entire 4-hr exposure period.

Observations made during the study included body weight, food consumption, breathing frequency, and startle reflex. After the last exposure, measurements and observations were made of the number of alveolar cells removed by lavage, clinical chemistry, pulmonary function, organ weights, and histopathology. An analysis of variance was conducted, and statistical significance was based on a significance level of 0.05.

Body Weight. No deaths occurred during this study, nor were there any overt clinical signs of toxicity. Sham-exposed animals had an initial weight loss during the exposure and then returned to normal weight. The exposed animals also had an initial weight loss upon exposure, but continued to lose weight until the beginning of the fourth week of exposure. After the fourth week, the animals started to gain weight. The males exposed to the lowest concentration gained weight more rapidly than those in the two higher concentrations; females gained little weight throughout the exposure period.

During the 2-month recovery period, females in all exposed groups grew more rapidly than males, so the end weights of the females were not different from those of the sham-exposed animals. On the other hand, males gained weight slowly, so only the lowest exposure group had body weights equal to those of the sham-exposed animals. Food consumption in rats of both sexes exposed to the lowest exposure concentration did not differ significantly from that in the sham-exposed group. In the two highest exposure groups, food consumption was less than that in the sham controls. Thus, in terms of weight loss, the aerosol concentrations used were above the no-observed-effect level.

Neurological Effects. A startle-reflex assay was used to test the time to reaction and the force of response to a sharp auditory stimulus. Rats were placed in a wire box within a larger sound-insulated box. A constant white noise at 85 decibels (dB) within the larger box helped eliminate outside noises. After an acclimation period of 10 min, rats received five 10-msec pulses of noise at 13,000 Hz and 110 dB separated by 25 sec. Their responses, or startle reflexes, were monitored by a Gould load cell under the wire box.

The number of responses made by exposed versus sham-exposed animals was not significantly different in any case. Reaction time (time from start of acoustic stimulus to start of response) was slightly longer in males examined 5 days after 13 weeks of exposure at the highest test concentration (1,600 mg/m^3). Female rats exposed at the middle test concentration (870 mg/m^3) also had increased reaction times. In both instances, the authors suggested that those differences, although statistically significant, were of doubtful biological importance because the changes were so small (less than 2 msec). The maximum amplitude of the response was statistically significantly higher in the sham-exposed controls than in other groups of males; recovery was complete 1 month after exposure. A similar effect was not observed in females. Statistically significant increases in peak time (time from acoustic stimulus to peak response) occurred in males, particularly in the group exposed at the highest concentration, for which every value measured at every time point through 2 months of recovery was higher than that in the sham-exposed controls. Males exposed at the two lowest concentrations also showed some statistically significant increases in peak time through 1 month of recovery. The differences varied up to 5 msec in females and 3 msec in males; the authors interpreted those values as a significant decrement in performance. In females, the increases in peak time were observed only in the two highest exposure groups, and the differences were no longer seen after 2 months of recovery.

The increases in time to peak response in these assays, in the absence of changes in reaction time and in the force exerted, represent a lengthening of the duration of the response. These changes are difficult to interpret. Most of the changes were approximately

2 or 3 msec, which the authors did not interpret as biologically significant. In the male rats, the peak times observed in the sham-exposed control rats differed from those in the rats in all other groups by 2.5 msec, even before any exposure occurred. This observation suggests that differences of 2 to 3 msec might not be biologically significant. Therefore, the subcommittee considered this response to be reversible and of doubtful biological concern.

Pulmonary Effects. Breathing frequencies measured 5 days after the last exposure were not affected. Some changes occurred in nitrogen washout curves, but the changes were not dose related. The nitrogen clearance rate was significantly faster in males and females exposed at the two lowest concentrations of the smoke than in the controls, but the group exposed at the highest concentration was marginally slower in its clearance rate than the controls. CO diffusing capacity did not change. Tests of obstructive airway disease, done using maximal forced exhalation maneuvers, showed that residual volume was significantly decreased at the highest exposure concentration. Functional residual capacity (FRC), vital capacity (VC), and peak flow were not affected. Total lung capacity (TLC) decreased slightly at the highest exposure concentration.

Total lavaged cells, as well as alveolar macrophages, showed some variable increases in the exposed animals, but by 2 months of recovery, differences between exposed and control groups in either males or females were not significant. The small increases seen in alveolar macrophages were not exposure-concentration dependent.

Organ Weights. Liver weights increased slightly in some of the exposed animals, but the increases were not concentration dependent. The same was true for adrenal weights. An increase was observed in wet weight and in wet-to-dry-weight ratio of the lung at the highest exposure concentration.

Clinical Chemistry. No biologically significant changes in the clinical chemistry values could be related to diesel-fuel exposure.

Histopathology. No microscopic lesions were recorded for the group that had been exposed at the highest concentration. The conclusions of the histopathological analysis were that no new lesions were found in any of the organs, including the respiratory tract, that could be attributed to inhalation of the diesel-fuel smoke.

Summary. Very little toxicity was observed in rats exposed for 13 weeks to diesel-fuel smoke at up to 1,600 mg/m^3.

Repeated Exposures: Relative Significance of Frequency, Duration, and Concentration

Using the same smoke-generation procedures as described by Dalbey and Lock (1982), Dalbey et al. (1982, 1987) tested the relative significance of frequency of exposure, duration of exposure, and aerosol concentration on resultant toxicity. Male and female Sprague-Dawley rats were exposed once a week for 9 weeks or three times a week for 3 weeks (a total of nine exposures in each case) to diesel-fuel smoke at concentrations ranging from 1,300 to 6,000 mg/m^3 for 2 or 6 hr. Examinations of animals were conducted within 2 days of the last exposure or after a 2-week recovery period using neurotoxicological tests, pulmonary-function tests, and evaluations of hematology, clinical chemistry, organ weights, and histopathology. A CT product of 8,000 $mg{\cdot}hr/m^3$ was used as a lower exposure concentration, and 12,000 $mg{\cdot}hr/m^3$ as an upper concentration, at which some deaths might be expected.

Neurological Effects. Dalbey et al. (1982, 1987) found no exposure-related changes in the neurotoxicity tests evaluated.

Pulmonary Effects. Dalbey et al. (1982, 1987) did not find histological abnormalities in any organ except the lung 2 days after the final exposure (Table 2-1). Focal accumulations of free pulmonary cells were observed in the lung parenchyma and were associated with thickening and hypercellularity of alveolar walls. The number of lavaged alveolar cells correlated well with histological observations, remaining elevated after the 2-week exposure period. Lung volumes were altered by exposure with an increased FRC, decreased TLC, and decreased VC. CO diffusing capacity was decreased in some of the exposed groups. Pulmonary wet weights (but not dry weights) increased with increasing exposures. The frequency of exposure appeared to be the dominant variable over

TABLE 2-1 Severity of Focal Pneumonitis and Associated Histological Changes Observed in the Lungs of Rats[a]

Exposures per Week	Hours per Exposure	Concentration (mg/m^3)	2 Days Post-Exposure	2 Weeks Post-Exposure
1	2	0	–	–
1	6	0	–	–
3	2	0	–	–
3	6	0	–	–
1	2	4,000	±	±
1	2	6,000	+	±
1	6	2,000	++	+
1	6	4,000/3,000	±	±/+
3	2	4,000	+/++	++
3	2	6,000	+++	+/++
3	6	2,000	++/+++	++
3	6	4,000/3,000	++/+++	±/+

[a] Severity was graded on the following relative scale: –, no observed focal pneumonitis; ±, equivocal change; +, slight; ++, moderate; +++, most severe observed. The total number of exposures in all cases was nine.

Sources: Dalbey and Lock 1982; Dalbey et al. 1987.

the range of variables studied; exposures three times per week were more injurious than once per week. Variation in duration of exposure appeared to have little effect, and dependence of the response on the exposure concentration was often lacking.

After a 2-week recovery period, the pulmonary effects observed 2 days after the final exposure were still present, but some were diminished. Alveolar macrophages continued to be approximately 2-fold higher in lavage fluid from exposed rats than from controls, but the neutrophils that were present in high numbers 2 days after exposure were now at control levels. Lung wet weights were still increased 2 weeks after the final exposure, but were slightly less than weights measured 2 days after the final exposure. The highest severity scores of the focal pneumonitis in the lung decreased from severe to moderate between 2 days and 2 weeks after exposure stopped.

Reproductive and Developmental Effects. Starke et al. (1987) conducted three types of studies to examine the potential reproductive and developmental effects of exposure of Sprague-Dawley rats to DF2 smoke: a prenatal developmental toxicity study, a dominant lethal study, and a single-generation study. Callahan et al. (1986) conducted a histological evaluation of reproductive organs following a 13-week exposure to diesel-fuel smoke. These studies are described below.

In the prenatal toxicity study, pregnant females were exposed on gestation days (GD) 6 to 15 to M60A1 tank-generated exhaust at 6 ± 6 mg/m^3 for 60 min (360 ± 360 mg·min/m^3), to DF2 smoke plus exhaust at 2,340 ± 450 mg/m^3 for 60 min (140,000 ± 27,000 mg·min/m^3), and to noise from the tank's engine for 20 to 30 sec (the time that the engine was run for the exhaust and smoke groups) (Starke et al. 1987). On GD 20, the females were killed and the fetuses examined for external, visceral, and skeletal malformations. Both the exhaust-exposed and the exhaust-DF2-exposed groups of females gained less weight between GD 6 and 20 than did the control females. The incidence of prenatal death was somewhat higher in the exhaust-DF2-exposed group (10%) than in the controls (2%), but it is within the range of the historical incidence of prenatal death in controls for this strain of rats. A cluster of five fetuses were malformed in one litter in the exhaust-DF2-exposed group, and one fetus was malformed in one litter in the exhaust-exposed group. The cluster is difficult to interpret, but because it was from a single litter, with no malformations from any other litter in that group, the subcommittee interprets this result as being unrelated to smoke exposure. The single malformation in the exhaust-exposed group is also unlikely to be related to the exhaust exposure.

The incidence of retarded or low skeletal ossification, particularly of the sternum, was higher in both the exhaust- and the exhaust-DF2-exposed groups (78% and 81%, respectively) than in the engine-noise-only control group (65%), but the difference was not statistically significant. The historical incidence of low ossification in control groups not exposed to engine noise is 7%. Thus, it appears that tank engine noise alone stresses the dams.

In the dominant lethal study, males that had sired viable offspring, referred to as proven males, were exposed for 10 weeks, 5 days per week, to noise control for 60 min, to tank exhaust for 15 or 60 min, and to exhaust plus DF2 smoke for 15 or 60 min (Starke et al. 1987). The exposure concentrations of tank exhaust and DF2 smoke were the same as those for the prenatal study. During the week following the exposure period, each male was housed with two unexposed females for 5 days, left alone for 2 days, then housed with two more females for 5 days. Females were killed 11 days after separation from the male. Starke et al. (1987) found no exposure-related differences in successful matings or in the incidence of single or multiple resorptions. Thus, no indications of dominant lethal mutations were observed.

In the single-generation study, proven males were exposed to the same conditions as those in the dominant lethal study (Starke et al. 1987). Females were exposed similarly but only during the last 3 weeks of the males' 10-week exposures. The week following the exposure, animals were mated (one male to two females), and daily exposure of the females continued through mating, gestation, and lactation; pups were not exposed directly. The pups were examined, sexed, and weighed 24 hr after birth. On postnatal day (PND) 4, they were reweighed and litters reduced to 10 pups each. On PND 21, pups were weighed and killed. Two males and two females were examined from each litter for external and visceral abnormalities. If no abnormalities were found, none of the remaining pups from that litter were examined. If abnormalities were found, the remaining pups from the litter were examined.

The pups from the 60-min exhaust-DF2-exposed group were reported to gain slightly less weight between days 1 and 21 than pups from the control group; however, the difference might be attributable to the slightly larger litter size of this group between days 1 and 4 (before standardizing the litter size) compared with the litter size of the control group. It is well known that average pup body weight tends to be inversely related to litter size, and that relationship was borne out in the comparison of the other exposure groups with the controls. For example, the 60-min exhaust-exposed group also had slightly larger litters than controls, and pup body-weight gain was slightly, but not significantly, smaller in this group than in controls. The 15-min exhaust- and exhaust-DF2-exposed groups had slightly smaller litters than controls, and pup weight gains were slightly, but not significantly, larger for these two groups than for controls. The incidence of malformations appeared to be similar among all the groups in the study; however, because not all pups were examined, these data are of limited value. In conclusion, the data from the single-generation study do not indicate any effect of either exhaust or exhaust-DF2-smoke exposure on the reproductive and developmental end points examined.

As described earlier, Callahan et al. (1986) exposed $B6C3F_1$ mice and F344 rats to an average of 2,300 mg/m^3 of M60A1 tank-generated exhaust-DF2 smoke for 15- or 60-min daily (5 days per week) for up to 13 weeks. No significant exposure-related changes in testes weight occurred in either species (other organ weights were not measured). Microscopic evaluation of the testes, mammary glands, ovary, and uterus also indicated no significant exposure-related lesions. However, adequate histological examination of the testis requires Bouin's fixation. The type of fixative used was not stated in the report, but formalin was likely used.

Thus, these studies provide no indication of a reproductive or developmental effect. The limitations of these studies include the following: (1) only one species was evaluated, (2) a second generation was not studied to determine the long-term effects of in utero and subsequent exposure, (3) a complete evaluation of adult male and female reproductive function was not conducted, and (4) only one test exposure concentration was used.

Screening Tests for Mutagenicity

A sample of diesel fuel containing 23.9% aromatics was negative for mutagenic potential in the *Salmonella* reversion and mouse lymphoma assays, both with and without metabolic activation (Conaway et al. 1982). Intraperitoneal (i.p.) injections of diesel fuel in rats was clastogenic in bone-marrow cells, inducing a significant 1% increase in chromosomal abnormalities at a dose of 2.0 mL/kg (but not at 0.6 mL/kg) (Conaway et al. 1982); however, the i.p. injection route might not be predictive of effects from inhalation exposures. DF2 concentrates were not found to be mutagenic in *Drosophila melanogaster* (Callahan et al. 1986).

Summary of Toxicity Data

The noncancer and carcinogenic effects of exposure to diesel-fuel smokes are summarized below. Table 2-2 summarizes the nonlethal dose-response data for inhalation exposures to diesel-fuel smokes.

Noncancer Toxicity

The high-concentration, one-time exposure studies of Dalbey

TABLE 2-2 Summary of Exposure-Response Data for Effects of Diesel-Fuel Smokes

Species, Smoke Type	Exposure Duration and Frequency	Concentration (mg/m^3)	Effects	Reference
Effects in Humans				
Human (workers)	10 min	170-330	No irritation	Il'in et al. 1969
Effects in Animals				
One-Time Inhalation Exposures				
Lethality				
Rat, mouse, guinea pig; DF1	15 min	15,000	No deaths	Callahan et al. 1983
	1 hr		Low mortality	
	2 hr		Moderate mortality	
	5 hr		All animals except 4 mice died	
Rat, mouse, guinea pig; DF2	15 min	35,000	Lethargy in rats and guinea pigs; no deaths	Callahan et al. 1983
Rat, mouse, guinea pig; DF2	1 hr	35,000	Sprague-Dawley rats: cyanotic F344 rats: 1/10 died Guinea pigs: 4/10 died	Callahan et al. 1983
Rat	2 hr	8,000	1/10 died	Dalbey and Lock 1982

Rat	2 hr	12,000	2/10 died	Dalbey and Lock 1982
Rat	2 hr	14,000	5/10 died	Dalbey and Lock 1982
Rat	2 hr	16,000	5/10 died	Dalbey and Lock 1982
Rat	4 hr	4,000	No deaths	Dalbey and Lock 1982
Rat	4 hr	6,000	No deaths	Dalbey and Lock 1982
Rat	4 hr	8,000	4/10 died	Dalbey and Lock 1982
Rat	4 hr	12,000	6/10 died	Dalbey and Lock 1982
Rat	4 hr	16,000	6/10 died	Dalbey and Lock 1982
Rat	6 hr	2,700	No deaths	Dalbey and Lock 1982
Rat	6 hr	4,000	No deaths	Dalbey and Lock 1982
Rat	6 hr	5,300	3/10 died	Dalbey and Lock 1982

Species, Smoke Type	Exposure Duration and Frequency	Concentration (mg/m^3)	Effects	Reference
Rat	6 hr	8,000	6/10 died	Dalbey and Lock 1982
Rat	6 hr	12,000	10/10 died	Dalbey and Lock 1982
Rat	6 hr	16,000	10/10 died	Dalbey and Lock 1982
Pulmonary Effects				
Rat, mouse, guinea pig; DF1	15 min, 1 hr, 2 hr, and 5 hr	15,000	Increasing severity of respiratory-tract lesions with increasing time of exposure; after 5 hr exposure, all animals except 4 mice died	Callahan et al. 1983
Repeated Inhalation Exposures				
Rat, mouse	15 min/d, 5 d/wk, 13 wk; 60 min/d, 5 d/wk, 13 wk	NOAEL 2,300	No observed adverse effects	Callahan et al. 1986

Rat	4 hr/d, 2 d/ wk, 13 wk	NOAEL 170	Decreased weight gain during exposures; females regained weight during recovery; males regained weight up to control values only at the lowest exposure concentration Startle reflex: Increased time to peak response in males at all exposures and in females at two higher exposures; no consistent treatment-related changes in reaction time or force exerted No histopathological or clinical chemistry changes; RV and TLC decreased at highest exposure, but no changes in FRC, VC, or peak flow	Lock et al. 1984
Rat	4 hr/d, 2 d/ wk, 13 wk	870		
Rat	4 hr/d, 2 d/ wk, 13 wk	1,600		
Rat	2 hr/d, 1 d/ wk, 9 wk	4,000	Equivocal FP at 2 d and 2 wk after exposure	Dalbey et al. 1982, 1987
Rat	2 hr/d, 1 d/ wk 9 wk	6,000	Slight FP at 2 d; equivocal at 2 wk	Dalbey et al. 1982, 1987
Rat	6 hr/d, 1 d/ wk, 9 wk	2,000	Moderate FP at 2 d; slight at 2 wk	Dalbey et al. 1982, 1987
Rat	6 hr/d, 1d/ wk, 9 wk	4,000	Equivocal FP at 2 d; equivocal to slight at 2 wk	Dalbey et al. 1982, 1987

Species, Smoke Type	Exposure Duration and Frequency	Concentration (mg/m³)	Effects	Reference
Rat	2 hr/d, 3 d/ wk, 3 wk	4,000	Moderate FP at 2 d; slight at 2 wk	Dalbey et al. 1982, 1987
Rat	2 hr/d, 3 d/ wk, 3 wk	6,000	Severe FP at 2 d; moderate at 2 wk	Dalbey et al. 1982, 1987
Rat	6 hr/d, 3 d/ wk, 3 wk	2,000	Severe FP at 2 d; moderate at 2 wk	Dalbey et al. 1982, 1987
Rat	6 hr/d, 3 d/ wk, 3 wk	4,000	Severe FP at 2 d; slight at 2 wk	Dalbey et al. 1982, 1987
Reproductive and Developmental Toxicity				
Rat, DF2	60 min/d, GD 6 to 15	NOAEL 2,340 ± 450	Prenatal toxicity: Slight increase in delayed skeletal development; same for exhaust-only and exhaust-smokes group	Starke et al. 1987
Rat (male), DF2	60 min/d, 5 d/wk, 10 wk	NOAEL 2,340 ± 450	Dominant lethal study: No observed adverse effects	Starke et al. 1987
Rat, DF2	60 min/d, 5 d/wk, ♂ 10 wk; ♀ 3 wk	NOAEL 2,340 ± 450	Single-generation study: No observed adverse effects	Starke et al. 1987

Rat, DF2	60 min/d, 5 d/wk, 13 wk	NOAEL 2,300	Subchronic study: No observed adverse effects on testes weight or histopathology of reproductive organs	Callahan et al. 1986

Abbreviations: hr, hour(s); d, day(s); wk, week(s); min, minute(s); FP, focal pneumonitis; GD, gestation day; ♂, male; ♀, female; DF2, diesel-fuel grade 2; DF1, diesel-fuel grade 1; NOAEL, no-observed-adverse-effects level.

and Lock (1982) and Callahan et al. (1983) indicated that the major target-organ system for diesel smoke via inhalation exposure is the respiratory tract. They found mortality from the smoke to be dependent on the product of exposure concentration and time (CT). Dalbey et al. (1982, 1987) found that the frequency of exposure also influences the toxicity of diesel-fuel smoke. Once-a-week exposures for 6 hr were much less toxic than 2-hr exposures, three times a week, to the same concentration of smoke. The study of Lock et al. (1984) was designed to determine if repeated exposures to low concentrations of diesel-fuel smoke would cause cumulative toxicity. Rats were exposed for 4 hr, twice a week, for 13 weeks, and observations were made 5 days and 2 months after the end of the exposures. The results indicated little toxicity at exposure concentrations of up to 1,600 mg/m^3, except for weight losses at all exposure concentrations, minimal pulmonary-function changes at the highest exposure concentration, and small reversible changes in time to peak startle response.

Carcinogenicity

Several tests for potential mutagenic activity of diesel fuels were negative (Conaway et al. 1982; Callahan et al. 1986). DF2 tested positive as a tumor promoter, but not as a complete carcinogen, in a mouse skin bioassay (Slaga et al. 1983). No long-term in vivo cancer bioassays have been done for diesel fuel per se. The appearance of tumors in mice exposed to some crude-oil-distillate fractions might be the result of the skin damage produced by repeated applications of the liquids to their shaved skin. Finally, IARC (1989) has concluded that there is inadequate evidence for the carcinogenicity of diesel fuels in humans, but limited evidence for the carcinogenicity of diesel fuel No. 4 (DF4) (also called marine diesel fuel), but not DF1 or DF2, in experimental animals.

Diesel exhaust (combusted diesel fuel) might account for 1% to 2% of diesel-fuel smoke if the smoke is generated by a diesel-powered vehicle. IARC has designated diesel exhaust as a 2B carcinogen (i.e., a possible human carcinogen with inadequate evidence in humans and adequate evidence in animals). However, the charge of the subcommittee was to review the toxicity of diesel-fuel smokes per se. The risk assessments for the diesel-fuel smoke were based on the concentration of particles of condensed fuel in the smoke, which, under obscuring conditions, are present at extremely high concentrations compared with soot from the diesel engines that are running the tanks.

LIMITS EXISTING RECOMMENDED EXPOSURE

Exposure limits have not been recommended for diesel-fuel smoke or its components. The ACGIH TLV-TWA value for diesel fuel of 350 mg/m^3 refers to total hydrocarbons as vapor, not as an aerosol of the total fuel (ACGIH 1995).

SUBCOMMITTEE EVALUATION AND RECOMMENDATIONS

On the basis of the toxicity information described above, the subcommittee has developed exposure guidance levels for military personnel exposed during an emergency release or during regular training exercises and for communities nearby training facilities to protect them from emergency or repeated releases of diesel-fuel smoke.

Inhalation studies indicate that the respiratory tract is the primary target organ system following exposure to diesel-fuel smoke and that mortality from a one-time exposure is dependent on C•T. Repeated-exposure studies indicate that pulmonary toxicity also depends on the frequency of exposure. One 6-hr exposure per week at the same concentration (4,000 mg/m^3) of smoke was less toxic to the lung than three 2-hr exposures a week (Dalbey et al. 1982, 1987). That finding indicates that for a given CT product, toxicity increases with the frequency of exposure. Therefore, the subcommittee recommends that training exercises without masking be conducted no more frequently than twice a week.

Military Exposures

Emergency Exposure Guidance Level (EEGL)1

For the EEGLs, the subcommittee considered the acute toxicity studies of Dalbey and Lock (1982) and Callahan et al. (1983). Dalbey and Lock (1982) used a probit analysis to estimate the CT product that induced 1% mortality. The 97.5% lower confidence bound of this value was 8,200 mg•hr/m^3. Considering the severity of the end point (death), the subcommittee divided that value by an uncertainty factor of 10 to predict a nonpermanent health impairment and by another uncertainty factor of 10 to account for interspecies sensitivity to arrive at an EEGL of 80 mg•hr/m^3. That CT product results in a 15-min EEGL of 320 mg/m^3, rounded to 300 mg/m^3, a 1-hr EEGL of 80 mg/m^3, and a 6-hr EEGL of 15 mg/m^3. Support for the assumption that permanent health effects would not occur from exposure at these concentrations comes from the repeated exposure studies of Dalbey et al. (1987), in which rats exposed once a week for 9 weeks to a weekly CT product of 8,000 mg•hr/m^3 exhibited only equivocal pulmonary inflammation, which remained equivocal after a 2-week recovery period. In addition, human volunteers exposed to diesel-fuel smoke at 330 mg/m^3 for 10 min reported no irritant effects.

Permissible Exposure Guidance Level (PEGL)2

The most relevant toxicity study for recommending PEGLs

[1] Guidance for a rare, emergency situation resulting in an exposure of military personnel.

[2] Guidance for repeated exposure of military personnel during training exercises.

involved repeated exposures of rats twice a week for 13 weeks (Lock et al. 1984). Exposure of rats for 4 hr, twice a week, to diesel-fuel smoke at concentrations as high as 1,600 mg/m^3 caused no lesions in the respiratory tract. The exposures did cause weight losses and reduced weight gains that resulted in weight deficits of greater than 10% for males, as compared with sham-exposed rats at the two highest exposure concentrations and with females even at the lowest exposure concentration. At the lowest exposure concentration, the weight deficit did not increase beyond 10% until after 10 weeks of exposure. Two months after the exposure, male body weights only returned to those of sham-exposed controls. Female weights in all exposure groups returned to control concentrations. If one considers the lower exposure concentration of 170 mg/m^3 for 4 hr, twice a week for at least 10 weeks, to be a no-observed-adverse-effect level (NOAEL, i.e., 1,360 mg•hr/m^3 per week), then the PEGL would be calculated as one-tenth that value (140 mg•hr/m^3 per week) to account for species differences.

The subcommittee also considered the studies of Dalbey et al. (1982, 1987). In those studies, rats exposed to 4,000 mg/m^3 of smoke for 2 hr, once a week for 9 weeks had equivocal signs of focal pneumonitis, and the equivocal lesion had not recovered 2 weeks later (Table 2-1). That regimen amounted to a weekly exposure of 8,000 mg•hr/m^3 and is considered a lowest-observed-adverse-effect level (LOAEL). Dividing by 100 to convert from the LOAEL to an expected NOAEL and to account for species differences, the PEGL is 80 mg•hr/m^3 within 1 week, with no more than two exposures per week.

The subcommittee chose 80 mg•hr/m^3 per week as the PEGL because it was the lower value of the two alternatives. The subcommittee assumed that an exposure event occurred over an extended period, such as 8 hr. Thus, the PEGL for a single exposure in 1 week would equal 10 mg/m^3 for an 8-hr exposure, and the PEGL for two exposures in 1 week would equal 5 mg/m^3 for each of two 8-hr exposures. The subcommittee recommends those PEGLs as ceiling values; in other words, those PEGLs apply even if the exposure events are less than 8 hr in a given day. The CT product for the PEGLs is the same as the CT product for the EEGLs; however, the maximum exposure concentration and rate of exposure are less for the PEGLs than for the EEGLs. The subcommittee also recommends that protective equipment be worn if exposures to diesel-fuel smoke during training appears to produce chronic dermatitis in any individuals.

Public Exposures

Short-Term Public Emergency Guidance Level (SPEGL)3

The SPEGL should accommodate the wide range of sensitivity possible in the general population compared with military personnel; therefore, a value equal to one tenth the EEGL is recommended (NRC 1986). It should be emphasized that the EEGL and SPEGL values are for emergency situations only. The SPEGL is 8 mg•hr/m^3 for a maximum of 6 hr for an emergency exposure.

Permissible Public Exposure Guidance Level (PPEGL)4

This value is equal to the PEGL divided by 10 to protect sensitive members of the public (NRC 1986). Thus, the subcommittee chose 8 mg•hr/m^3 per week as the PPEGL, with no more than two exposures per week. As for the PEGL, the subcommittee assumed that an exposure event occurs over an extended period, such as 8 hr. Thus, the PPEGL for a single exposure in 1 week would equal 1 mg/m^3 for an 8-hr exposure, and the PPEGL for two exposures in 1 week would equal 0.5 mg/m^3 for each of two 8-hr

[3] Guidance for a rare, emergency situation potentially resulting in an exposure of the public to a military-training smoke.

[4] Guidance for repeated exposures of public communities near military-training facilities.

exposures. The PPEGLs represent ceiling values; in other words, those PPEGLs apply even if the exposure events are less than 8 hr in a given day. The CT product for the PPEGLs is the same as the CT product for the SPEGLs; however, the maximum exposure concentration and rate of exposure are less for the PPEGLs than for the SPEGLs.

Summary of Subcommittee Recommendations

The exposure guidance levels for diesel-fuel smoke for military personnel are summarized in Table 2-3. The exposure guidance levels for diesel-fuel smoke to protect the public in the vicinity of training facilities are summarized in Table 2-4.

TABLE 2-3 EEGLs and PEGLs for Diesel-Fuel Smoke for Military Personnel

Exposure Guideline	Exposure Duration	Guidance Level (mg/m^3)
EEGL	15 min	300
	1 hr	80
	6 hr	15
PEGL	8 hr/d, 1 d/wk	10
	8 hr/d, 2 d/wk	5

TABLE 2-4 SPEGLs and PPEGLs for Diesel-Fuel Smoke at the Boundaries of Military Training Facilities

Exposure Guideline	Exposure Duration	Guidance Level (mg/m^3)
SPEGL	15 min	30
	1 hr	8.0
	6 hr	1.5
PPEGL	8 hr/d, 1 d/wk	1
	8 hr/d, 2 d/wk	0.5

RESEARCH NEEDS

Recommendations for diesel-fuel smoke were based on studies conducted in laboratory animals over relatively short durations, all less than 3 months. It is not known if the modest adverse effects noted at the end of these exposures would have increased in severity had the exposures continued for longer durations. Some military personnel are exposed to smoke during training exercises over several years. Toxicity studies conducted over exposure durations longer than 10 weeks, perhaps up to 1 or 2 years, would provide the information necessary to evaluate human health risks due to years of exposure to diesel-fuel smoke during training exercises.

Army personnel who work with this smoke, trainers in particular, represent a rich source of potential information on the health effects of the smoke. The subcommittee recommends that the U.S. Army conduct a prospective study with appropriate controls in which pulmonary-function tests (spirometry and diffusing capacity at a minimum) and routine chemistry tests (panel 20 plus Mg and thyroid tests as a minimum) are conducted on personnel who are exposed repeatedly to the smoke.

Additional studies are recommended to explore fully the potential for reproductive and developmental effects of exposure to diesel-fuel smokes. To evaluate male and female reproductive toxicity, a two-generation study, including a detailed evaluation of reproductive effects, is recommended. Developmental effects have been evaluated only in Sprague-Dawley rats; conclusions would be more robust if similar findings were observed in a second species of mammal.

REFERENCES

ACGIH (American Conference of Governmental Industrial Hygienists). 1995. 1995-1996 Threshold Limit Values and Biological Exposure Indices. American Conference of Governmental Industrial Hygienists, Cincinnati, Ohio.

Blackburn, G.R., R.A. Deitch, C.A. Schreiner, M.A. Mehlman, and

C.R. Mackerer. 1984. Estimation of the dermal carcinogenic activity of petroleum fractions using a modified Ames assay. Cell Biol. Toxicol. 1:67-80.

Beck, L.S., D.I. Hepler, and K.L. Hansen. 1982. The acute toxicology of selected petroleum hydrocarbons. Pp. 1-12 in Proceedings of the Symposium: The Toxicology of Petroleum Hydrocarbons , N.H. MacFarland, C.E. Holdsworth, J.A. MacGregor, R.W. Call, and M.L. Kane, eds. Washington, D.C.: American Petroleum Institute.

Callahan, J.F., C.L. Crouse, G.E. Affleck, R.L. Farrand, R.W. Dorsey, M.S. Ghumman, R.J. Pellerin, D.H. Heitkamp, C. Lilly, J.J. Feeney and J.T. Weimer. 1983. The Acute Inhalation Toxicity of Diesel Fuels (DF2 and DF1) Used in Vehicle Engine Exhaust Smoke Systems (VEESS). Tech. Rep. ARCSL-TR-82064. Chemical Systems Laboratory, U.S. Army Armament, Munitions and Chemical Command, Aberdeen Proving Ground, Edgewood, Md.

Callahan, J.F., C.L. Crouse, G.E. Affleck, E.G. Cummings, R.L. Farrand, R.W. Dorsey, M.S. Ghumman, R.D. Armstrong, W.C. Starke, R.J. Pellerin, D.C. Burnett, D.H. Heitkamp, C. Lilly, J.J. Feeney, M. Rausa, E.H. Kandel, J.D. Bergmann, and J.T. Weimer. 1986. The Subchronic Inhalation Toxicity of DF2 (Diesel Fuel) Used in Vehicle Engine Exhaust Smoke Systems (VEESS). Tech. Rep. CRDCTR-85009. Chemical Research and Development Center, U.S. Army Armament, Munitions and Chemical Command, Aberdeen Proving Ground , Edgewood, Md.

Conaway, C.C., C.A. Schreiner, and S.T. Cragg. 1982. Mutagenicity evaluation of petroleum hydrocarbons. Pp. 128-138 in Proceedings of the Symposium: The Toxicology of Petroleum Hydrocarbons, N.H. MacFarland, C.E. Holdsworth, J.A. MacGregor, R.W. Call, and M.L. Kane, eds. Washington, D.C.: American Petroleum Institute.

Dalbey, W., and S. Lock. 1982. Chemical Characterization and Toxicological Evaluation of Airborne Mixtures. Inhalation Toxicology of Diesel Fuel Obscurant Aerosol in Sprague-Dawley Rats, Final Report, Phase 1, Acute Exposures. ORNL/TM-8867. AD-A132 650. Oak Ridge National Laboratory, Oak Ridge, Tenn.

Dalbey, W., S. Lock, and R. Schmoyer. 1982. Chemical Characterization and Toxicological Evaluation of Airborne Mixtures. Inhalation Toxicology of Diesel Fuel Obscurant Aerosol in Sprague-Dawley Rats, Final Report, Phase 2, Repeated Exposures. ORNL/TM-9169.

AD-A142 540. Oak Ridge National Laboratory, Oak Ridge, Tenn.

Dalbey W., M. Henry, R. Holmberg, J. Moneyhun, R. Schmoyer, and S. Lock. 1987. Role of exposure parameters in toxicity of aerosolized diesel fuel in the rat. J. Appl. Toxicol. 7:265-275.

Dautrebande, L., and R. Capps. 1950. Studies on aerosols. IX. Enhancement of irritating effects of various substances on the eye, nose, and throat by particulate matter and liquid aerosols in connection with pollution of the atmosphere. Arch. Int. Pharmacodynam. Ther. 82:505-528.

IARC (International Agency for Research on Cancer). 1989. Diesel fuels. Pp. 219-237 in IARC Monographs on the Evaluation of Carcinogenic Risks to Humans: Occupational Exposures in Petroleum Refining; Crude Oil and Major Petroleum Fuels, Vol. 45. Lyon, France: International Agency for Research on Cancer.

Il'in, B.I., L.I. Kogan, and N.V. Buzulutskii. 1969. Suppurative diseases in persons working with fuels and lubricants [in Russian]. Voen. Med. Zh. 9:69.

Ingram, A.J., and P. Grasso. 1991. Evidence for and possible mechanisms of non-genotoxic carcinogenesis in mouse skin. Mutat. Res. 248:333-340.

Jenkins, R.A., R.W. Holmberg, J.S. Wike, J.H. Moneyhan, and R.S. Brazell. 1983a. Chemical Characterizations and Toxicologic Evaluation of Airborne Mixtures. ORNL/TM-9196. AD-A142 718. Oak Ridge National Laboratory, Oak Ridge, Tenn.

Jenkins, R.A., D.L. Manning, M.P. Maskatinec, J.H. Moneyhun, and W. Dalbey. 1983b. Chemical Characterization and Toxicologic Evaluation of Airborne Mixtures. Diesel Fuel Smoke Particulate Dosimetry in Sprague-Dawley Rats. ORNL/TM-9195. AD-A142 914. Oak Ridge National Laboratory, Oak Ridge, Tenn.

Lewis, S.C., R.W. King, S.T. Cragg, and D.W. Hillman. 1984. Skin carcinogenic potential of petroleum hydrocarbons: Crude oil, distillate fractions and chemical class subfractions. Pp. 139-150 in Advances in Modern Environmental Toxicology: Applied Toxicology of Petroleum Hydrocarbons, Vol. 6, N.H. MacFarland, C.E. Holdsworth, J.A. MacGregor, R.W. Call, and M.L. Kane, eds. Princeton, N.J.: Princeton Scientific Publishers.

Liss-Suter, D., R. Mason, and P.N. Craig. 1978. A Literature Review—Problem Definition Studies on Selected Toxic Chemicals: Occupational Health and Safety Aspects of Diesel Fuel and White Smoke

Generated From It, Vol. 1. DAMD17-77-C-7020. AD-A056 018. Science Information Services Department, Franklin Institute Research Laboratories, Rockville, Md.

Lock, S., W. Dalbey, R. Schmoyer, and R. Griesemer. 1984. Chemical Characterization and Toxicological Evaluation of Airborne Mixtures. Inhalation Toxicology of Diesel Fuel Obscurant Aerosol in Sprague-Dawley Rats, Final Report, Phase 3, Subchronic Exposures. ORNL/TM-9403. AD-A150 100. Oak Ridge National Laboratory, Oak Ridge, Tenn.

NRC (National Research Council). 1986. Criteria and Methods for Preparing Emergency Exposure Guidance Level (EEGL), Short-term Public Emergency Guidance Level (SPEGL), and Continuous Exposure Guidance Level (CEGL) Documents. Washington, D.C.: National Academy Press.

Slaga, T.J., L.L. Triplett, and R.J.M. Fry. 1983. Chemical Characterization and Toxicological Evaluation of Airborne Mixtures. Tumorigenicity Studies of Diesel Fuel-2, Red Smoke Dye, and Violet Smoke Dyes in the SENCAR Mouse Skin Tumorigenesis Bioassay System. ORNL/TM-9752. AD-A159 728. Oak Ridge National Laboratory, Oak Ridge, Tenn.

Starke, W.C., R.J. Pellerin, D.C. Burnett, J.H. Manthei, and D.H. Heitkamp. 1987. Teratogenicity, Mutagenicity, and Effects of Grade 2 Diesel Fuel on Reproduction in a Single Generation of Rats. Tech. Rep. CRDEC-TR-87083. Chemical Research, Development and Engineering Center, U.S. Army Armament, Munitions and Chemical Command, Aberdeen Proving Ground, Edgewood, Md.

3

Fog-Oil Smoke

BACKGROUND INFORMATION

Military Applications

Fog-oil smoke is the term used to describe an oil smoke generated by injecting mineral oil into a heated manifold. The oil vaporizes upon heating and condenses when exposed to the atmosphere, producing respirable particles. Troops are exposed to fog-oil smoke when it is used as a visual obscurant during training or in combat. Graphite can be added to fog oil to provide screening in the infrared range of the electromagnetic spectrum. Fog oil used without graphite is evaluated in this chapter.

Physical and Chemical Properties

Composition:	Variable; see description below
Minimum flash point:	160°C
Viscosity, kinematic, centistokes:	3.40 to 4.17 at 100°C
Pour point:	-40°C
Boiling point:	300 to 600°C

To meet military specifications (for pour point[1] and cloud point[2]), fog oil historically has been produced from naphthenic oils. The composition varies from batch to batch from different sources and even from the same source. Samples taken from two sources of conventionally refined fog oil contained approximately 50% aromatic hydrocarbons, 1% acids, alcohols and esters, and nitrogen derivatives in the parts-per-million range (Katz et al. 1980). Only slight variation in chemical composition results from the smoke-generation process (Katz et al. 1980). The severely refined fog oil that is in use today should not contain detectable quantities of many aromatic hydrocarbons.

Fog-oil smoke is a condensation aerosol (a mist) composed of liquid particles. Condensation aerosols are, in general, relatively small in aerodynamic size and respirable and are generated to obscure soldiers from view. A fraction of the oil (components with low boiling points) might remain in the vapor form. All measurements of fog-oil smoke reported or recommended in this chapter are referred to in milligrams of total particulates per cubic meter.

The chemical and physical properties of fog oil are similar to those of lubricating and petroleum-based cutting oils. Substances added to cutting and lubricating oils to maintain their physical properties during use under extreme pressure and heat are responsible for the distinguishing characteristics of these oils. Although little information is available regarding the health effects of fog oil, inferences can be drawn to a large extent from the health effects of lubricating and mineral oils. However, only certain cutting oils would be appropriate in making such a comparison. Insoluble cutting oils composed of mineral oils with only small quantities of additives should have biological properties similar to fog oil.

---

[1] The lowest temperature at which a liquid will flow when its container is inverted.
[2] The temperature at which a waxy solid material appears as the liquid is cooled.

Emulsified cutting oils have a greater complexity of additives than insoluble cutting oils, and synthetic cutting oils contain no mineral oil. Thus, those oils cannot be compared validly with fog oil.

Droplets of cutting- and lubricating-oil mists found in occupational settings also are largely in the respirable range (Jones 1961). Increased incidences of skin cancer of the hands, arms, and scrotum have been observed in workers exposed to conventionally refined mineral oils in the jute, cotton-spinning, and metal-machining industries. Polycyclic aromatic hydrocarbons (PAHs) and related heterocyclic compounds are thought to be responsible for these effects (Bingham et al. 1965; Halder et al. 1984; IARC 1984; Kane et al. 1984).

In this report, the subcommittee distinguishes between "old" and "new" fog oils. Old fog oils (basically, naphthenic oils) are similar to conventionally refined mineral oils, which contain various carcinogenic or potentially carcinogenic substances, including PAHs and related heterocyclic compounds. The military specification for fog oil was changed after IARC (1984) concluded that untreated naphthenic oils were carcinogenic. The new military specification for fog oil excludes all carcinogenic or potentially carcinogenic constituents. Fog oil procured after the new specification was implemented in 1986 is referred to as new fog oil. Industry uses two processes to remove carcinogenic and potential carcinogenic constituents: (1) severe solvent refining or extraction removes undesirable compounds, and (2) hydro-treatment converts them to less toxic saturated compounds.

Military Exposures

Young et al. (1989) collected 1-hr personal[3] and area air samples during training sessions at the U.S. Army Chemical School at

[3] Personal air samples were taken by a device worn on the lapel and were used to measure ambient concentrations in the breathing zone of an individual.

Fort McClellan. Students were learning operations and maintenance (O&M) for smoke generators and participating in field training exercises (FTX) in three courses: advanced individual training (54B10AIT), the basic noncommissioned officers' course (BNCOC), and the commissioned officers' basic course (COBC). Personal samples were taken for both cadre (O&M only) and students. Fog-oil-smoke exposure concentrations ranged from 0 to 680 mg/m^3. The mean exposures were much higher for O&M training than for FTX (i.e., 69 ± 10 mg•hr/m^3 versus 8.7 ± 1.3 mg•hr/m^3). The O&M training took place for 4 hr each day for 2 days. Unlike FTX, students and cadre are required to stand near the smoke generators to make adjustments for the entire training session. The difference between students' and cadres' exposure concentrations was not statistically significant. Both students and cadre typically are exposed for a total of 8 hr in a 2-day course. However, the cadre teach many courses over a 3- to 4-year assignment at the Chemical School. Thus, the potential for chronic exposure is much greater for cadre than for students.

Advanced individual training led to higher fog-oil-smoke exposures than the basic courses for both FTX and O&M. Area-sample-concentration measurements did not differ significantly from the personal-sample measurements taken simultaneously.

The mass median aerodynamic diameter (MMAD) of the fog-oil-smoke particles ranged between 1 and 3 μm (Young et al. 1989). During one laboratory test, old-fog-oil smoke gave MMADs of 2.43 and 2.21 μm, with geometric standard deviations of 1.68 μm and 1.64 μm, respectively (Cataldo et al. 1989). In a study of smoke dispersion at Eglin Air Force Base, Florida, the mean diameter (presumably by count) of fog-oil-smoke particles ranged from 0.505 to 2.10 μm (Policastro and Dunn 1985). Because the count mean diameter should be less than the MMAD for a given distribution of particle sizes, that measurement is in rough agreement with the measurements of Young et al. (1989) and Cataldo et al. (1989). Thus, a large portion of the fog-oil-smoke particle mass is in the respirable-size range.

Studies of fog-oil-smoke dispersion were conducted at Dugway

Proving Ground in 1985 (Liljegren et al. 1988). Fog-oil-smoke plumes were produced using M3A3E3 smoke generators, and samples were taken from 25 to 800 m downwind to measure concentration, particle-size distribution, deposition on surfaces, and chemical composition. Concentration measurements, taken 25 m downwind of the source along the fog-oil plume centerline, were presented for only three experiments. The highest concentrations—120, 110, and 33 mg/m^3—were observed at the sampling site nearest the generator. Two hundred meters from the source, two of the centerline concentrations were just above the detection limit (1 mg/m^3) and one was below the detection limit. Farther than a few hundred meters downwind, the concentrations exhibited considerable spatial heterogeneity.

Liljegren et al. (1988) found that particle size was distributed log-normally and that the MMAD was about 0.7 μm. The chemical compositions of the raw oil, the initial smoke particles, and the smoke particles at the farthest point from the generator were not detectably different, and deposition of smoke particles on vertical and horizontal surfaces was not statistically significant.

TOXICOKINETICS

No data are available to evaluate the toxicokinetics of fog-oil smoke or aerosols of similar oils in humans or in animals.

TOXICITY SUMMARY

Effects in Humans

Dermal Exposures

Noncancerous Skin Lesions

Short exposures to lubricating oils can cause mild erythema. More prolonged exposure can cause inflammation, dermatitis, folliculitis, acne, eczema, and contact sensitivity (Cruickshank and Gourevitch 1952). Those effects have been reported for conventionally refined oils (i.e., those not having undergone severe solvent refining or hydro-treatment). The PAH content of those oils is thought to be responsible for those conditions. Support for that theory comes from skin-painting studies in animals, which show that highly refined mineral oils (similar to new fog oil) are not likely to cause serious chronic skin conditions (Bingham et al. 1965; Bingham and Horton 1966). Dermatitis, folliculitis, and warts have been reported in men exposed to poorly refined cutting and lubricating oils (Cruickshank and Squire 1950; Hodgson 1970).

Cancer of the Skin and Scrotum

There is ample evidence pointing to an association between exposure to conventionally refined mineral oils and skin and scrotal cancer (Bingham et al. 1980; IARC 1984). IARC (1984) concluded that evidence was sufficient to consider conventionally refined mineral oils to be carcinogenic to humans. Tumors of the skin of the scrotum, arms, and hands are a result of sprays from the machines and direct contact with oil-coated surfaces, particularly along the lower abdominal area (Cruikshank and Squire 1950; Cruikshank and Gourevitch 1952). Chronic inflammatory and cancerous lesions on the hands, forearms, and scrotum developed in 60% of workers exposed to liquid cutting lubricants for over 15 years at their jobs in the United Kingdom (Hodgson 1973). Case-control studies of Connecticut workers exposed to cutting oils demonstrated excess sinonasal and scrotal cancers (Roush et al. 1980, 1982). Benzo(*a*)pyrene and other PAHs in lubricating oils were identified in a region of France in which a high incidence of skin cancer was observed (Thony et al. 1975, 1976). In jute and cotton textile workers exposed to high concentrations of mineral oils, high rates of skin and scrotal cancer have been noted (Kinnear et al. 1954).

Scrotal cancer is uncommon in men not exposed to mineral

oil (Hodgson 1973). The incidence of scrotal and skin cancer in textile workers and machinists appears to have declined in recent years. That decrease has been attributed to the new refining processes that reduce the PAH content of oils (Falk et al. 1964; Bingham et al. 1980). Thus, exposure to new fog oil would not represent a major concern for skin and scrotal cancer.

Multiple Routes of Exposure

In addition to inhalation exposures in occupational settings, workers can be exposed to oil mists that settle on equipment, skin, and clothing, thereby causing dermal and oral exposures. The primary health effects associated with occupational exposures to oil mists include pulmonary effects and skin cancer.

Pulmonary Effects

Pulmonary effects, such as granulomas or pneumonias, can occur with exposure to highly refined mineral oils that lack PAHs. Over 400 cases of lipoid pneumonia resulting from ingestion, inhalation of oil-based nose drops, or intralaryngeal injection of medicinal oil were reported in the literature before 1978 (IARC 1984).

Lipoid pneumonia can be of two types: (1) lipoid granuloma or paraffinoma, which is a local lesion within a single lobe of the lung, and (2) diffuse pneumonitis, which is characterized by oil droplets that are widely disseminated throughout one or more lobes of the lung. Fibrosis can result from lipoid pneumonia, leading to loss of lung function (Proudfit et al. 1950; Jampolis et al. 1953).

Lipoid pneumonia, however, is rarely seen in the workplace even when concentrations of oil mists are over 50 mg/m^3 (Liss-Suter et al. 1978). A survey conducted by the American Petroleum Institute of workers exposed to mineral-oil mists showed no instances in which lung abnormalities were associated with oil exposure. The threshold for discomfort seems to be 5 mg/m^3 (Hendricks et al. 1962). In these studies, the average exposure was 15 mg/m^3, with measurements ranging from 1 to 57 mg/m^3. On the basis of these studies, Hendricks et al. (1962) recommended a maximum allowable exposure level of 5 mg/m^3 to avoid nuisance and subjective complaints.

In a study by Jones (1961), 19 workers from a steel-rolling mill were examined after having been exposed for 9 to 18 years to oil-mist concentrations as high as 9 mg/m^3 for 2 hr per day, 5 days per week. The oil was a naphthenic spindle oil containing petroleum sulfonates, rosin soap, and cresylic acid. No respiratory diseases were noted, nor were any skin or gastric disorders observed; however, an increase in linear striations were discovered in the lungs of 12 men. Jones (1961) concluded that the importance of that finding was not known.

Persistent minor respiratory-tract infections were evident in workers exposed to mineral-water emulsions resulting in oil-mist concentrations averaging 2 mg/m^3. However, the symptoms could not be associated with occupational exposure to the mist (Hervin and Lucas 1972). Excess respiratory symptoms (cough and phlegm) were noted in nonsmoking and smoking machine-shop workers exposed to median oil-mist concentrations of 3.2 to 4.5 mg/m^3 for at least 3 years (Jarvholm et al. 1982). The reported incidence of chronic cough and phlegm was higher for the more-exposed workers in grinding and hardening than for the less-exposed workers employed in the turning department; however, those symptoms might have been due to the additives in the oils (Jarvholm et al. 1982). Lung function (1-sec forced expiratory volume (FEV_1), forced vital capacity (FVC), residual volume, closing volume, and diffusion capacity) was not impaired in the nonsmokers examined (lung function was not evaluated for smokers) (Jarvholm et al. 1982).

Ely et al. (1970) found that oil-mist concentrations of about 1 mg/m^3 (median) to 5.2 mg/m^3 (mean) did not result in any abnormalities in the incidence of cough, bronchitis, wheeze, and dyspnea or in FEV_1 and FVC in machinists exposed for 8 hr per day, 5 days per week, for 1 to 38 years (mean 13 years). Individual measurements with air sampled for at least 1 hr ranged from 0.07 to 110 mg/m^3.

There have been case reports attributing occupational exposure to oil mists as the causative factor in respiratory illness. One subject with lipoid pneumonia, chronic cough, frequent colds, and substantial loss of pulmonary function was reported by Proudfit et al. (1950).

Greaves et al. (1997a,b) examined a group of 1,882 automobile workers composed of machinists exposed to aerosols from metal-working fluids and unexposed assemblers at three plants. The metal-working fluids were either straight mineral oils, soluble-oil emulsions, or synthetic fluids. Average exposure of the three unexposed groups (assemblers) was 0.10 to 0.15 mg/m^3, expressed as "thoracic" aerosol fraction, and average exposure of the three exposed groups (machinists) was 0.16 to 0.80 mg/m^3. Individual exposures were 0.07 to 0.44 mg/m^3 for the assemblers and 0.16 to 2.43 mg/m^3 for the machinists. The machinists had all been exposed for at least 6 months, and a majority had been exposed for over 2 years.

A respiratory questionnaire and lung spirometry were used to determine the effects of exposure. The straight oils, which would be most similar to fog oil, produced respiratory symptoms of phlegm and wheezing as well as chest tightness and breathlessness. Those effects were greater than those observed for the soluble-oil group and less than those observed for the synthetic-oil group (Greaves et al. 1997a,b). Lung spirometry demonstrated a greater effect on FEV_1 than on FVC. The results were consistent with an obstructive ventilatory function and were evident at the highest exposure concentrations of straight and soluble oils (Greaves et al. 1997a,b). However, both the straight and soluble oils could have included up to 40% additives, which might have been responsible for the observed pulmonary effects.

Drasche et al. (1974) evaluated respiratory questionnaires

completed by German workers exposed to drilling- and cutting-oil mists at concentrations of 40 to 150 mg/m^3 for long periods of time, and found no indications of respiratory irritation that could be attributed to the oil-mist exposures. However, Drasche et al. (1974) did not report how the air concentrations were measured or how the questionnaires were administered.

Skyberg et al. (1986) found an increased prevalence of slight basal-cell lung fibrosis in workers exposed to oil mists and kerosene vapors compared with workers in the same company not exposed to those substances. Eight-hour time-weighted average (TWA) oil-mist concentrations measured by personal air samplers ranged from 0.15 to 0.3 mg/m^3 among the exposed workers. However, most exposure would occur during short intervals when workers cleaned oil-containing pans (one to three daily) and cleaned large vessels from the inside (two to four times a month). A peak concentration of 2,000 to 4,000 mg/m^3 was measured one time over a pan-cleaning operation (total number of area measurements not reported); no measurements were taken in the enclosed vessels entered by workers for cleaning. Calibration of the GF/A glass-fiber personal air monitors against Millipore HA membrane filters indicated that the personal air monitors underestimated ambient oil-mist concentrations by at least 20-fold. Thus, it is likely that those workers were exposed repeatedly to relatively high concentrations of oil mists for short periods of time and that actual 8-hr TWA exposures were higher than 5 mg/m^3 for some workers.

In considering all the data reporting respiratory effects of oil mists, isolated cases of adverse effects (dyspnea, bronchitis, wheeze, fibrosis, and impaired pulmonary function) resulted from occupational and nonoccupational exposures. The majority of studies, however, do not point to serious respiratory problems from concentrations commonly found in industrial settings, and the problems that have been reported could be due to the additives used to maintain the oils' physical characteristics under high pressure and temperature.

Carcinogenic and Mutagenic Effects

Excesses of lung cancer in oil-exposed workers have been observed in certain studies (Coggon et al. 1984; Vena et al. 1985), and other studies have been negative (Decoufle 1978; Jarvholm et al. 1981). In Kodak plants in New York State, exposures to oil mist in concentrations ranging from 0.07 to 110 mg/m^3 (median concentration, 1.5 mg/m^3; mean concentration, 3.7 mg/m^3) demonstrated no excess deaths from cancers at all sites combined or from respiratory tract cancer, Hodgkins disease, or leukemia (Ely et al. 1970). Waterhouse (1971) found a significant excess of primary cancers of the respiratory and upper digestive tracts in men with mineral-oil-related cancers of the scrotum. This study examined the records of primary cases of scrotal cancer in the Birmingham Regional Cancer Registry for 1950 to 1967. In a cohort study of men exposed to synthetic, emulsified, and insoluble cutting oils, an excess of gastrointestinal-tract cancer, but not respiratory-tract cancer, was found (Decoufle 1976, 1978). Due to the mixed exposures, the relevance of these results to mineral oils, including fog oil, can be questioned. Cancer mortality in a large number of workers in various Japanese industries demonstrated an association between gastric cancer and machine-oil exposure (Okubo and Tsuchiya 1974). Bell et al. (1987) determined that the risk for malignant melanoma was significantly increased in workers exposed to cutting oils but not in those exposed to mineral oils. They concluded that the risk for melanoma was probably due to nitrosamines in the cutting oils.

Meaningful conclusions cannot be drawn from most of the studies linking cancers of other organs to mineral oil. Most of the studies provide no information regarding exposure concentrations or the chemical composition of the oils.

Peripheral lymphocytes cultured from pressed-glass makers exposed to mineral-oil mists with relatively high concentrations of PAHs had a significantly higher frequency of aberrant cells and chromosome breaks per cell. The exposure concentration was less than 5 mg/m^3, and the exposure duration was not reported (Sram et al. 1985).

Effects in Animals

Dermal Exposures

Lethality

Lethality of old fog oil in rabbits by single dermal application was more than 2 g/kg (Mayhew et al. 1985), indicating low potential for acute toxicity from dermal exposures.

Skin Irritation

Slight-to-moderate irritation was produced by a single application of mineral oils to the skin of rabbits (Beck et al. 1982; Mayhew et al. 1985). Repeated application, however, can cause more damage. Marked epidermal hypertrophy, hyperplasia, hyperkeratosis, and depilation were produced when conventionally refined light mineral oil was applied every other day for 1 week to the skin of guinea pigs. Those effects were produced by nonaromatic as well as aromatic compounds. Hydrocarbons with carbon numbers from C_{14} to C_{19} caused greater damage than those with higher carbon numbers (C_{21} to C_{23}) (Hoekstra and Phillips 1963).

Cancer

Several skin-painting studies have shown that conventionally refined mineral oils are carcinogenic via dermal exposures (Bingham et al. 1965; Jepsen et al. 1977). Severe solvent extraction or hydro-treatment reduces or eliminates the tumorigenicity of mineral oils (Bingham and Horton 1966; Kane et al. 1984). The PAH content of the oils is thought to be responsible for the tumorigenicity (Bingham et al. 1965). Used oils, which are likely to have more PAHs due to the formation of pyrolysis products, tend to be more tumorigenic (Jepsen et al. 1977), although that is not always the case (Stemmer and King 1979).

Cragg et al. (1985) found that the *Salmonella*/microsome assay did not predict the dermal carcinogenic activity of complex petroleum mixtures; the mixtures were not mutagenic in the bacterial bioassay but were slightly to highly carcinogenic in mouse skin. McKee et al. (1989) found that repeated dermal application of diesel fuel, which contains low levels of biologically active PAHs, caused tumors in mouse skin. They suggested that the chronic irritation and hyperplasia produced by the diesel fuel act as tumor promoters. Similar considerations might apply to the tumorigenicity of mineral oils, although they are less irritating than diesel fuel. Mineral oils have been found to contain co-carcinogens, tumor promoters, and tumor antagonists, which more than likely account for the lack of direct correlation between PAH content and tumorigenicity (Roe et al. 1967; Bingham et al. 1980).

A histological examination of mouse skin removed from mice after exposure to oil for 3 days showed enlargement of nuclei. The enlargement correlated well with the carcinogenicity of the oils as demonstrated by mouse-skin-painting studies (Ingram and Grasso 1987).

Oral Exposures

One-Time Exposures

Lethality of old fog oil and paraffinic and naphthenic lubricating oils in rats by single oral intubation was more than 5 g/kg (Beck et al. 1982), indicating little potential for acute lethality with oral administration.

Repeated Exposures

In one study using dietary administration of 2% liquid paraffin (comparable to medicinal-grade mineral oil) for 500 days and in another study using three grades of 5% petrolatum for 2 years, no oil-related tumors were found (Schmal and Reiter 1953; Oser et al. 1965).

Inhalation Exposures

One-Time Exposures

Lethality. A steep dose-response curve was found for single exposures to old fog oil when inhaled by rats. After 3.5 hr of exposure at a concentration of 1,000 mg/m^3, no animals died, but after 6 hr of exposure at the same concentration, 20% of the exposed animals died. The concentration estimated to kill 50% of test organisms (LC_{50}) after a 3.5-hr exposure was 5,200 mg/m^3. After the same exposure duration, less than 15% of the animals died at 4,000 mg/m^3 and over 80% died at 6,000 mg/m^3(Grose et al. 1986; Selgrade et al. 1987) (see Table 3-1).

Pulmonary Effects. Submicron mists of medicinal-grade mineral oil, laboratory-grade paraffin oil, grade S-75 light lubricating oil, and SAE 10W-30 motor oil were tested on guinea pigs. Single 1-hr exposures at concentrations of 10 and 40 mg/m^3 produced alterations in pulmonary function (Costa and Amdur 1979). Light lubricating oil at a concentration of 200 mg/m^3, however, caused a decrease in pulmonary compliance (Costa and Amdur 1979). Shoshkes et al. (1950) found that 2-hr exposures of mice to animal, vegetable, mineral, and SAE No. 10 motor oil at atmospheric concentrations of 4,500 mg/m^3 caused only the appearance of scattered macrophages in lung tissue.

TABLE 3-1 Acute Lethality of Old-Fog-Oil Smoke via Inhalation Exposure

Species	Exposure Duration	Exposure Concentration (mg/m^3)	End Point and Comments	Reference
Rat	6 hr	1,000	20% died	Grose et al. 1986
Rat	3.5 hr	1,000	No mortality	Selgrade et al. 1987
		4,000	<15% died	
		5,200	LC_{50}	
		6,000	> 80% died	

Repeated Exposures

Lethality. Two of six monkeys exposed to mists of SAE No. 10 automobile lubricating oil at a concentration of 132 mg/m^3 (30 min per hr, 24 hr per day, for up to 100 days) died within 100 days (Lushbaugh et al. 1950). However, six of seven monkeys exposed to mists of SGF No. 1 diesel lubricating oil at 63 mg/m^3 (for up to 1 year) died, indicating a higher toxicity of SGF No. 1 compared with SAE No. 10 oils (Lushbaugh et al. 1950).

Pulmonary Effects. Four-week exposures of mice to mineral-oil aerosols at a concentration of 4,500 mg/m^3 caused localized foreign-body reactions and lipoid pneumonia (Shoshkes et al. 1950). Edible oils had no effects.

Monkeys and CF_1 mice were exposed to mists of SAE No. 10 automobile lubricating oil at concentrations of 132 mg/m^3 for 30 min per hr, 24 hr per day, for up to 100 days. Rats, rabbits, monkeys, and strain-A mice were exposed to SGF No. 1 diesel-engine lubricating oil at concentrations of 63 mg/m^3 for up to 1 year, presumably for 30 min per hr, 24 hr per day, 7 days per week. Small amounts of oil were retained in the lungs of mice, rats, and rabbits, and macrophages with dispersed small oil droplets were seen.

Increases in macrophages were seen initially, but remained constant from week 5 to the end of the study (Lushbaugh et al. 1950).

The effects were more severe in monkeys, which retained a larger amount of the oils in their lungs. As indicated above, SGF No. 1 oil was more toxic (killing six of seven animals tested) than SAE No. 10 (killing two of six within 100 days). Infectious pneumonitis, pulmonary lipophages, and severe hyperplastic gastritis were observed in animals exposed to both oils. Increases in the number of alveolar macrophages, diffuse pneumonitis, and acute pneumonia with edema were among the other effects observed. Longer exposures resulted in diffuse acute bronchopneumonia with edema and hemorrhage, lobular pneumonia, diffuse pneumonitis, and fibroplasia. Notwithstanding the respiratory effects, the cause of death for most monkeys was severe hyperplastic gastritis.

Grose et al. (1986) exposed rats to old fog oil at 500 and 1,500 mg/m^3 for 70 min or 3.5 hr per day for 2 or 4 days per week for 4 weeks. Dose-related accumulation of alveolar macrophages and wet and dry lung weights were elevated in animals exposed at 1,500 mg/m^3 for 70 min or 3.5 hr for both 2 and 4 days. In all groups of animals exposed at 1,500 mg/m^3, total lung protein, total cell count, and polymorphonuclear leukocytes were elevated in bronchoalveolar lavage fluid. The animals also exhibited mild inflammatory pulmonary edema. End-expiratory volume was elevated 21% in high-dose animals, and no significant changes occurred in low-dose animals.

Rats exposed to old fog oil at 500 and 1,500 mg/m^3 for 4 hr per day, 4 days per week, for 13 weeks demonstrated concentration-dependent accumulation of macrophages in alveoli and peribronchial lymph nodes. The effects persisted after a 4-week recovery period. Male rats exposed to 1,500 mg/m^3 showed focal hemorrhage and multifocal granulomatous pneumonia. The appearance of the granulomas 4 weeks after exposure suggests development of a progressive lesion (Grose et al. 1986).

In a final study, Grose et al. (1986) exposed rats to old fog oil at 200 and 500 mg/m^3 for 3.5 hr per day, 4 days per week, for 13 weeks. The increase in alveolar macrophages was slight to moderate at 500 mg/m^3 and minimal to slight at 200 mg/m^3. Increases in dry and wet lung weight were not significant at 200 mg/m^3 but were significant at 500 mg/m^3. Bronchial alveolar lavage protein and liver aryl hydrocarbon hydroxylase (AHH) activity increased, but a decrease was observed in zoxazolamine-induced paralysis time. It was thought that the AHH increase was due to the PAH in the fog oil (Grose et al. 1986).

The series of studies by Grose et al. (1986) demonstrated adverse effects down to 200 mg/m^3 after 13 weeks of exposure. Thus, a no-observed-adverse-effect level (NOAEL) was not determined. Four- and 13-week exposures of male and female rats to old-fog-oil aerosols caused inflammatory edema, but pulmonary function and gas exchange were not significantly affected. Granulomas in rats exposed to 500 and 1,500 mg/m^3 persisted through the 4-week recovery period, suggesting a progressive lesion in the lung after subchronic exposure.

Wagner et al. (1964) exposed five species—the rat, rabbit, dog, hamster, and mouse—to white mineral-oil mists at 5 or 100 mg/m^3 for varying periods of 6 months to 2 years. Mice were exposed daily for 6 hr at 5 mg/m^3 for 12 months or at 100 mg/m^3 for 16 months. For dogs, 6, 12, or 26 months of daily 6-hr exposures were conducted for the 5- and 100-mg/m^3 concentrations. Rabbits were exposed at 5 mg/m^3 daily for 6 or 12 months or at 100 mg/m^3 daily for 6, 12, or 18 months. Rats and hamsters were exposed similarly to rabbits, the exception being that the rat and hamster 100-mg/m^3 groups were sacrificed at 16 and 15 months, respectively.

The white mineral oil is comparable to new fog oil. Rats and dogs were most affected by the oil. Exposures at 100 mg/m^3 caused pulmonary lipoid granulomas in the dog and pneumonitis in the rat. No pathological effects were found at 5 mg/m^3. Based on the study by Wagner et al. (1964), the 5-mg/m^3 concentration could represent a NOAEL. Because serum alkaline phosphatase levels correlated well with histopathological findings in the dog, rat, and rabbit, Wagner et al. (1964) concluded that alkaline phosphatase could be used as an indicator of early injury from pulmonary irritants.

Gastrointestinal Effects. In the Lushbaugh et al. (1950) experiments with monkeys exposed to mists of SAE No. 10 automobile lubricating oil and SGF No. 1 diesel lubricating oil (described above), the cause of death of most monkeys was severe hyperplastic gastritis.

Carcinogenic Effects. Exposure for 1 year to SGF No. 1 diesel-engine lubricating oil at 63 mg/m^3 did not cause tumors in strain-A mice (Lushbaugh et al. 1950); neither did exposure of CAF_1/Jax mice to mineral oil (comparable to new fog oil) at 5 and 100 mg/m^3 for 13 months (Wagner et al. 1964). Both strains are highly susceptible to the development of lung tumors.

Reproductive and Developmental Toxicity

No data are available on the reproductive and developmental toxicity of fog oil in mammals by any exposure route.

Screening Tests for Carcinogenicity and Mutagenicity

A naphthenic-based lubricating oil stock similar in viscosity to fog oil was negative in the L5178-Y-mouse-lymphoma assay and did not cause chromosomal aberrations in the rat bone-marrow cytogenetics assay (Conaway et al. 1984).

Using a modification of the Ames assay, Blackburn et al. (1986) determined that the mutagenicity of 18 oil samples correlated well with their tumorigenic potency in mouse-skin-painting studies. Skisak et al. (1987) used this modified assay to test 26 distillation fractions and found a high correlation between mutagenic activity and the tumorigenic potency demonstrated in mouse-skin-painting assays. Severe hydrogenation decreases the mutagenicity of lubricating-oil base stocks (Venier et al. 1987). Solvent refining also decreases the mutagenicity of these stocks (Hermann et al. 1980a, b). New fog oil was negative in the *Salmonella* Ames assay (Lee et al. 1989).

Summary of Toxicity Data

Table 3-1 (above) summarized the acute lethality of inhalation exposure of rats to old-fog-oil aerosols. Table 3-2 summarizes the available exposure-response data for nonlethal effects in humans and animals of exposures to aerosols of fog oil and similar mineral oils. The type of oil aerosol to which the humans or animals were exposed is indicated in the first column of the table.

Noncancer Toxicity

Dermal exposures to oils that are similar to old fog oils can produce adverse effects, including skin inflammation, dermatitis, and folliculitis, but short-term dermal exposures to the oils that are similar to new fog oil are unlikely to produce more than temporary irritation.

The acute lethality of exposure to old fog oil is low. One report from Grose et al. (1986) indicates an inhalation LC_{50} of 5,200 mg/m^3 in rats exposed for 3.5 hr to old fog oil. Similarly, reports of lethality from either oral (lethal dose to 50% of the test animals, or $LD_{50} > 5{,}000$ mg/kg) or dermal ($LD_{50} > 2{,}000$ mg/kg) exposures of animals indicate that old fog oil is relatively nontoxic following short-term exposures.

The respiratory tract is a primary target organ for exposure to fog-oil aerosols. A single 2-hr exposure of mice to conventionally refined mineral oil at 4,500 mg/m^3 produced scattered macrophages in lung tissue, but no other pulmonary effects (Shoshkes et

TABLE 3-2 Summary of Exposure-Response Data for Nonlethal Effects of Exposure to Aerosols of Fog Oil and Similar Oils

Receptor; Type of Oil	Exposure Frequency and Duration	NOAEL (mg/m^3)	LOAEL (mg/m^3)	End Point and Comments	Reference
Effects in Humans					
Repeated Exposures					
Pulmonary Effects					
Workers; mineral oil	Not reported	2 (mean)	—	No adverse pulmonary effects	Hervin and Lucas 1972
Workers; mineral oil	Not reported	W 5	—	Threshold for "discomfort"	Hendricks et al. 1962
Workers; mineral oil	Not reported	15 (mean)	—	No adverse pulmonary effects	Hendricks et al. 1962
Workers; napthenic spindle oil	2 hr/d, 5 d/wk, 9 to 18 yr	up to 9 (mean)	—	Increased striations in lungs of 12/19 men; significance unknown	Jones 1961
Nonsmoking machinists; various oils	8 hr/d, 5 d/wk, 1 to 38 yr (mean 13 yr)	1.0 to 5.2[a]	—	No respiratory effects	Ely et al. 1970

Receptor; Type of Oil	Exposure Frequency and Duration	NOAEL (mg/m^3)	LOAEL (mg/m^3)	End Point and Comments	Reference
Machinists (smoking and nonsmoking); various mineral oils	8 hr/d, 5 d/wk, at least 3 yr	2.0^b (median)	3.2^b (median)	Increasing chronic cough and phlegm	Jarvholm et al. 1982
Machinists (smoking and nonsmoking); various mineral oils	8 hr/d, 5 d/wk, at least 3 yr	4.5 (median)	—	No effects on measures of pulmonary function	Jarvholm et al. 1982
Machinists; straight mineral oils, with from 0% to 40% additives	8 hr/d, 5 d/wk, D 6 mo	—	0.16 to 2.43^c	Phlegm, wheezing, chest tightness, breathlessness, obstructive ventilatory function	Greaves et al. 1997a,b
Workers; drilling and cutting oils	Not reported	40 to 150	—	No respiratory effects; methods not reported	Drasche et al. 1974
Carcinogenic Effects	Workers in various industrial settings (exposures not quantified) experienced elevated skin and scrotum cancer associated with exposures to conventionally refined mineral oils (like old fog oil)				
Kodak plant workers; various oils	8 hr/d, 5 d/wk, > 5 yr	3.7 (mean)	—	No excess deaths from cancer at any site	Ely et al. 1970

Effects in Animals					
One-Time Exposures					
Pulmonary Effects					
Guinea pig; various oils	1 hr	–	10	Altered pulmonary function	Costa and Amdur 1979
Guinea pig; various oils	1 hr	–	200	Decreased pulmonary compliance	Costa and Amdur 1979
Mouse; mineral oil	2 hr	–	4,500	Scattered macrophages in lung tissue	Shoshkes et al. 1950
Repeated Exposures					
Lethality					
Monkey; SGF No. 1 oil	30 min/ hr, 24 hr/ d, 100 d	–	132	6/7 animals died	Lushbaugh et al. 1950
Monkey; SAE No. 10 oil	30 min/ hr, 24 hr/ d, 100 d	–	132	2/6 died within 100 d	Lushbaugh et al. 1950
Pulmonary Effects					
Mouse; mineral oil	4 wk	–	4,500	Lipoid pneumonia	Shoshkes et al. 1950
Monkey; SGF No. 1 and SAE No. 10 oils	30 min/ hr, 24 hr/ d, up to 100 d	–	132	Infections pneumonitis, pulmonary lipophages, edema, other pulmonary effects	Lushbaugh et al. 1950

Receptor; Type of Oil	Exposure Frequency and Duration	NOAEL (mg/m^3)	LOAEL (mg/m^3)	End Point and Comments	Reference
Rat; old fog oil	70 or 210 min/d, 2 or 4 d/wk, 4 wk	500	1,500	Elevated total lung protein, cell count, and pmn leukocytes in lavage fluid; mild inflammatory edema	Grose et al. 1986
Rat; old fog oil	4 hr/d, 4 d/wk, 13 wk	–	500	Accumulation of macrophages in alveoli and peribronchial lymph nodes	Grose et al. 1986
Male rat; old fog oil	4 hr/d, 4 d/wk, 13 wk	500	1,500	Granulomatous pneumonia after 4 wk exposure	Grose et al. 1986
Rat; old fog oil	3.5 hr/d, 4 d/wk, 13 wk	200	500	Increase in alveolar macrophages; changes in lung weight	Grose et al. 1986
Mouse; white mineral oil	6 hr/d, 7 d/wk, 16 mo	100	–	No adverse effects	Wagner et al. 1964
Rat; white mineral oil	6 hr/d, 7 d/wk, 6 to 16 mo	5	100	Pneumonitis	Wagner et al. 1964

Dog; white mineral oil	6 hr/d, 7 d/ wk, 6 to 24 mo	5	100	Pulmonary lipoid granulomas	Wagner et al. 1964
Rabbit; white mineral oil	6 hr/d, 7 d/ wk, 18 mo	100	–	No adverse effects	Wagner et al. 1964
Hamster; white mineral oil	6 hr/d, 7 d/ wk, 6 to 15 mo	5	100	Increase in basic alkaline phosphatase and magnesium-activated alkaline phosphatase	Wagner et al. 1964
Gastrointestinal Effects					
Monkey; SGF No. 1 and SAE No. 10 oils	30 min/ hr, 24 hr/d, up to 100 d	–	63	Hyperplastic gastritis	Lushbaugh et al. 1950
Carcinogenic Effects					
Strain A mouse; SGF No. 1 oil	30 min/ hr, 24 hr/d, 1 yr	63	–	No tumors	Lushbaugh et al. 1950
CAF_1-Jax mouse; white mineral oil	6 hr/d, 7 d/ wk, 13 mo	100	–	No tumors	Wagner et al. 1964

Abbreviations: hr, hour(s); min, minute(s); d, day(s); wk, week(s); mo, month(s); yr, year(s); NOAEL, no-observed-adverse-effect level; LOAEL, lowest-observed-adverse-effect level; pmn, polymorphonuclear.

[a] The time-weighted average exposure was likely to be somewhere between the median (1.0 mg/ m^3) and mean (5.2 mg/m^3) exposure concentrations recorded.

[b] The authors believed that the increased cough and phlegm could be due to the additives in the oil.

[c] The observed effects could have resulted from the additives in the oils.

al. 1950). Studies of dogs, rabbits, rats, hamsters, and mice exposed to white mineral oils for 1 to 2 years identified a NOAEL for repeated exposures of 5 mg/m^3 and a lowest-observed-adverse-effect level (LOAEL) of 100 mg/m^3 for pulmonary effects (Wagner et al. 1964). Repeated-exposure concentrations higher than that can result in a variety of adverse pulmonary effects in animals.

Adequate human data are not available to establish NOAEL values for adverse health effects resulting from short-term exposures. For long-term exposures, Hendricks et al. (1962) found no illnesses related to inhalation of oil aerosols in a considerable population of individuals occupationally exposed for long periods. Long-term average exposure concentrations were 15 mg/m^3. Concentrations in individual air samples ranged from 1 to 57 mg/m^3; at concentrations less than 5 mg/m^3, few complaints were noted.

Carcinogenicity

Conventionally refined mineral oils, which are chemically similar to old fog oil, have been shown to cause cancer of the skin of the arms, hands, and scrotum of humans. IARC (1984) recognizes eight classes of mineral oils based on increasing severity of processing or refinement. New fog oil might correspond roughly to the more severely refined oils included in IARC class 4, which are hydro-treated oils, or to the less severely refined oils in class 5, such as analytic-grade white mineral oils. IARC (1984) stated that there is sufficient evidence that mildly hydro-treated mineral oils in class 4 are carcinogenic to experimental animals, but the available data on severely hydro-treated oils in class 4 are inadequate to permit an evaluation of their carcinogenicity to experimental animals. IARC (1984) stated further that the combination of hydro-treatment and solvent extraction appears to reduce or eliminate skin tumorigenicity of mineral oils. Moreover, there is no evidence of tumorogenicity of analytic-grade white mineral oils (class 5) by any route of administration except by intraperitoneal injection; the relevance of that route of exposure is unclear. Thus, fog oil that has undergone hydro-treatment and solvent extraction is unlikely to be carcinogenic via inhalation or dermal exposures.

EXISTING RECOMMENDED EXPOSURE LIMITS

The U.S. Occupational Safety and Health Administration (OSHA) and the American Conference of Governmental Industrial Hygienists (ACGIH) established a Threshold Limit Value (TLV) TWA of 5 mg/m^3 for exposures to severely refined mineral-oil mists of 8 hr per day, 5 days per week, based on the studies by Hendricks et al. (1962) and Wagner et al. (1964) (ACGIH 1991). ACGIH (1995) recently proposed further that the sum total of PAHs listed as carcinogenic by the National Toxicology Program (NTP) should not exceed 5 $\mu g/m^3$. Palmer (1990) reviewed the effects of exposure to fog-oil smoke in humans and animals and recommended that the U.S. Army adopt the ACGIH TLV-TWA of 5 mg/m^3 for new fog oil.

SUBCOMMITTEE EVALUATION AND RECOMMENDATIONS

The subcommittee endorses the recommendations of Palmer (1990) to ensure that carcinogenic compounds are not in the fog oil used by the military and also recommends that representative batches of the oil be analyzed for the 15 PAHs listed as carcinogens by the NTP. The subcommittee also recommends exposure guidance levels for new fog oil, as described below.

Type of Fog Oil

Conventionally refined mineral oils, which are similar to old fog oil, have been shown to cause cancer of the skin of the arms, hands, and scrotum of humans, whereas mineral oils are thought unlikely to be carcinogenic after undergoing severe solvent refining or hydro-treatment. To protect military personnel from cancer risks associated with old fog oil, the exposure limits would be so low that it would require continuous masking in the vicinity of fog oil. Prohibiting the use of old fog oil might be more feasible than recommending an exposure limit for it. Given the impracticality of donning appropriate protective apparel in most situations in which fog oil would be used and given the fact that the oil can penetrate ordinary military uniforms, the subcommittee endorses prohibiting the use of old fog oil for production of smokes to which military personnel would be exposed.

In some cases, severe solvent refining or hydro-treatment does not remove all carcinogens. Thus, the subcommittee also endorses a military specification for fog oil that requires manufacturer testing of the oil to ensure the absence of carcinogenic constituents. The subcommittee recommends that manufacturers use the modified Ames test of Blackburn et al. (1986) and the Food and Drug Administration (FDA 1979) test for white-oil purity.

Finally, the subcommittee endorses testing the current inventory of fog oil purchased after the specifications were revised by using the FDA test for white-oil purity (FDA 1979) and the modified Ames test of Blackburn et al. (1986) to ensure that all batches are not carcinogenic.

Military Exposures

The recommendations in this section are made with the assumption that exposures of military personnel are to smokes generated from new rather than old fog oil.

Emergency Exposure Guidance Level (EEGL)4

The potential for death of animals from a one-time exposure to fog oil is low. One report from Grose et al. (1986) indicated an

[4] Guidance for a rare, emergency situation resulting in an exposure of military personnel.

LC_{50} of 5,200 mg/m^3 in rats exposed for 3.5 hr to old fog oil. Other reports of lethality from either oral (LD_{50} > 5 g/kg) or dermal (LD_{50} > 2 g/kg) exposures also indicated that old fog oil is relatively nontoxic following short-term exposures. No acute toxicity information is available for humans. Existing guidelines for occupational exposures (e.g., those of ACGIH and the Michigan and Detroit Bureaus of Industrial Hygiene for 40-hr work weeks) have been set at 5 mg/m^3 to avoid complaints by workers.

Given the lack of data on health effects of short-term exposures of humans to mineral-oil mists, the subcommittee used the LOAEL for pulmonary effects in mice exposed for 2 hr at 4,500 mg/m^3 (Shoshkes et al. 1950) as the point of departure for estimating EEGLs. The subcommittee divided the NOAEL by a factor of 10 to estimate effects in humans from data on animals and by another factor of 10 to estimate a NOAEL from a LOAEL. To estimate exposure guidance levels for exposure durations less than 2 hr, Haber's law was applied based on the similarity of fog-oil and diesel-fuel smokes (both petroleum based). Data for diesel-fuel smoke indicate that C•T is a good predictor of mortality (see Chapter 2). Applying Haber's law to the 2-hr exposure guidance level of 45 mg/m^3 resulted in a 15-min EEGL of 360 mg/m^3, a 1-hr EEGL of 90 mg/m^3, and a 6-hr EEGL of 15 mg/m^3. The Hendricks et al. (1962) study, which indicated that no adverse health effects in workers were associated with 8-hr exposures to an average of 15 mg/m^3 mineral-oil mists, also supports a 6-hr EEGL of at least 15 mg/m^3. The subcommittee believes that it is reasonable for the 15-min EEGL of 360 mg/m^3 for fog-oil smoke to be higher than the 15-min EEGL of 300 mg/m^3 for diesel-fuel smoke, because new fog oil contains essentially no aromatic hydrocarbons, whereas diesel fuel is approximately 15% aromatics, and because new fog oil is less irritating to the skin than is diesel fuel. The aromatic compounds are thought to contribute to the acute toxicity of these petroleum products. The fog-oil EEGL might be more conservative than necessary (i.e., it could be higher than 360 mg/m^3) , and the subcommittee recommends that the U.S. Army conduct research to identify a more accurate value.

Permissible Exposure Guidance Levels (PEGL)5

In the Hendricks et al. (1962) study in which a large number of individuals exposed to oil mists were investigated, the lack of illness related to inhalation of the mists was striking. Exposure concentrations ranged from 1 to 57 mg/m^3, averaging 15 mg/m^3. At concentrations less than 5 mg/m^3, few or no complaints were noted. On the basis of these studies, the subcommittee developed an 8-hr, 5 days per week, PEGL of 5 mg/m^3. Exposures during training exercises often exceed the recommended PEGL (Liljegren et al. 1988; Young et al. 1989). Thus, careful adherence to respiratory protection policy is recommended.

Public Exposures to New-Fog-Oil Smoke

The recommendations in this section are made with the assumption that the military is using new rather than old fog oil.

Short-Term Public Emergency Guidance Level (SPEGL)6

Although the possibility is slight that a short-term public-health emergency would occur from new-fog-oil-smoke exposure, general discomfort might occur at concentrations above 5 mg/m^3 (Hendricks et al. 1962). No serious effects have been reported in humans working in industrial atmospheres with concentrations averaging as high as 15 mg/m^3; however, the subcommittee believes that a lower SPEGL would be appropriate to protect sensitive subpopulations that might be exposed. Therefore, the subcommittee

[5] Guidance for repeated exposure of military personnel during training exercises.

[6] Guidance for a rare, emergency situation potentially resulting in an exposure of the public to a military-training smoke.

recommends dividing the EEGLs by an uncertainty factor of 10 to account for the potentially greater range in sensitivities of the general public compared with industrial workers. Thus, the 15-min SPEGL is 36 mg/m^3, the 1-hr SPEGL is 9.0 mg/m^3, and the 6-hr SPEGL is 1.5 mg/m^3.

Permissible Public Exposure Guidance Level (PPEGL)[7]

There is a dearth of information on the effects of exposure to old or new fog oil on reproduction and development as well as on sensitive populations. Thus, the subcommittee recommends that a safety factor of 10 be applied to the PEGL to estimate the PPEGL. The resultant PPEGL is 0.5 mg/m^3.

Summary of Subcommittee Recommendations

Table 3-3 summarizes the subcommittee's recommended exposure guidance levels for exposure of military personnel to new-fog-oil smoke. Table 3-4 summarizes the subcommittee's recommended exposure guidance levels for new-fog-oil smoke for the boundaries of military-training facilities to ensure that public communities near the training facilities are not at risk of adverse effects.

RESEARCH NEEDS

There is no information regarding the health effects of short-term (i.e., from a few minutes to a few hours) exposure of humans to fog-oil or severely refined mineral-oil aerosols at concentrations above 15 to 60 mg/m^3. Moreover, there are no human or animal

[7] Guidance for repeated exposures of public communities near military-training facilities.

data that can be used to evaluate the extent to which Haber's law applies to health effects of these oils. The development of dangerously low visibility and slippery surfaces might occur at concentrations less than those that could impair human performance as a result of toxic effects or physical impairment of pulmonary function; however, no data are available to evaluate that possibility. Thus, studies on the health effects of short-term exposures that also evaluate the applicability of Haber's law are needed to provide more sound guidance for emergency exposures.

TABLE 3-3 EEGLs and PEGL for New-Fog-Oil Smoke for Military Personnel

Exposure Guideline	Exposure Duration	Guidance Level (mg/m^3)
EEGL	15 min	360
	1 hr	90
	6 hr	15
PEGL	8 hr/d, 5 d/wk	5

TABLE 3-4 SPEGLs and PPEGL for New-Fog-Oil Smoke at the Boundaries of Military Training Facilities

Exposure Guideline	Exposure Duration	Guidance Level (mg/m^3)
SPEGL	15 min	36
	1 hr	9.0
	6 hr	1.5
PPEGL	8 hr/d, 5 d/wk	0.5

The information available from studies of occupational exposure of humans is insufficient to rule out the possibility of long-term health effects. Moreover, few animal studies are available to evaluate the long-term health effects of repeated exposures to new-fog-oil smoke at concentrations of 5 to 60 mg/m^3. Thus, long-term, repeated-exposure studies should be conducted by using appropriate small mammal species and exposure levels likely to be experienced by military personnel in the field.

Information is not available on reproductive and developmental toxicity in mammals. Increasing numbers of females are recruited into the military. Thus, studies should also be conducted to ascertain reproductive developmental toxicity in mammals. These studies should use the inhalation route if possible.

To ensure protection of the public, some effort is warranted to determine whether some human subpopulations might be more sensitive than others. Short-term exposure of individuals with and without asthma at the SPEGL, followed by pulmonary-function tests (spirometry and diffusing capacity as a minimum requirement), could be useful both in determining whether those with asthma are more sensitive and whether the SPEGL is adequate or overprotective.

Finally, the subcommittee notes that Army personnel who work with this smoke, trainers in particular, are potentially a rich source of information on the health effects of the smoke. The subcommittee recommends that the Army conduct a prospective study with appropriate controls in which pulmonary-function tests and routine chemistry tests (panel 20 plus Mg and thyroid tests as a minimum requirement) are performed on personnel who are exposed repeatedly to the smoke.

REFERENCES

ACGIH (American Conference of Governmental Industrial Hygienists). 1991. Documentation of the Threshold Limit Values and Biological Exposure Indices, 6th Ed. American Conference of Governmental Industrial Hygienists, Cincinnati, Ohio.

ACGIH (American Conference of Governmental Industrial Hygienists). 1995. 1995-1996 Threshold Limit Values and Biological Exposure Indices. American Conference of Governmental Industrial Hygienists, Cincinnati, Ohio.

Beck, L.S., D.I. Hepler, and K.L. Hansen. 1982. The acute toxicology of selected petroleum hydrocarbons. Pp. 1-12 in Proceedings of the Symposium: The Toxicology of Petroleum Hydrocarbons, N.H. MacFarland, C.E. Holdsworth, J.A. MacGregor, R.W. Call, and M.L. Kane, eds. Washington, D.C.: American Petroleum Institute.

Bell, C.M.J., C.M. Jenkinson, T.J. Murrells, R.G. Skeet, and J.D. Everall. 1987. Aetiological factors in cutaneous malignant melanomas seen at a UK skin clinic. J. Epidemiol. Community Health 41:306-311.

Bingham, E., and A.W. Horton. 1966. Environmental carcinogenesis: Experimental observations related to occupational cancer. Adv. Biol. Skin 7:183-193.

Bingham, E., A.W. Horton, and R. Tye. 1965. The carcinogenic potency of certain oils . Arch. Environ. Health 10:449-451.

Bingham, E., R.P. Trosset, and D. Warshawsky. 1980. Carcinogenic potential of petroleum hydrocarbons. A critical review of the literature. J. Environ. Pathol. Toxicol. 3:483-563.

Blackburn, G.R., R.A. Deitch, C.A. Schreiner, and C.R. Mackerer. 1986. Predicting carcinogenicity of petroleum distillation fractions using a *Salmonella* mutagenicity assay. Cell Biol. Toxicol. 2:63-84.

Cataldo, D.A., P. Van Voris, M.W. Ligotke, R.J. Fellow, B.D. McVeety, S.-m.W. Li, H. Bolton, Jr., and J.K. Frederickson. 1989. Evaluate and Characterize Mechanisms Controlling Transport, Fate and Effects of Army Smokes in an Aerosol Wind Tunnel: Transport, Transformations, Fate and Terrestrial Ecological Effects of Fog Oil Obscurant Smokes. AD-A20414. Pacific Northwest Laboratory, Richland, Wash.

Coggon, D., B. Pannett, and E.D. Acheson. 1984. Use of job-exposure matrix in an occupational analysis of lung and bladder cancers on the basis of death certificates. J. Natl. Cancer Inst. 72:61-65.

Conaway, C.C., C.A. Schreiner, and S.T. Cragg. 1984. Mutagenicity evaluation of petroleum hydrocarbons. Pp. 89-107 in Advances in Modern Environmental Toxicology, Vol. 6. Applied Toxicology of Petroleum Hydrocarbons, M.A. Mehlman, ed. Princeton, N.J.: Princeton Scientific Publishers.

Costa, D.L., and M.O. Amdur. 1979. Respiratory response of guinea pigs to oil mists. Am. Ind. Hyg. Assoc. J. 40:673-679.

Cragg, S.T., C.C. Conaway, and J.A. MacGregor. 1985. Lack of concordance of the *Salmonella/* microsome assay with the mouse dermal carcinogenesis bioassay for complex petroleum hydrocarbon mixtures. Fundam. Appl. Toxicol. 5:382-390.

Cruickshank, C.N.D., and A. Gourevitch. 1952. Skin cancer of the hand and forearm. Br. J. Ind. Med. 9:74-79.

Cruickshank, C.N.D., and J.R. Squire. 1950. Skin cancer in the engineering industry from the use of mineral oil. Br. J. Ind. Med. 7:1-11.

Decoufle, P. 1976. Cancer mortality among workers exposed to cutting-oil mist. Ann. N.Y. Acad. Sci. 271:94-101.

Decoufle, P. 1978. Further analysis of cancer mortality patterns among workers exposed to cutting oil mists. J. Natl. Cancer Inst. 61:1025-1030.

Drasche, H., L. Finzel, H. Martschei, and R. Meyer. 1974. Industrial-medical investigations of persons exposed to oil mists [in German]. Zentralbl. Arbeitsmed. 24:305-312.

Ely, T.S., S.F. Pedley, F.T. Hearne, and W.T. Stille. 1970. A study of mortality, symptoms, and respiratory function in humans occupationally exposed to oil mist. J. Occup. Med. 12:253-261.

Falk, H.L., P. Kotin, and A. Mehler. 1964. Polycyclic hydrocarbons as carcinogens for man. Arch. Environ. Health 8:721-729.

FDA (Food and Drug Administration). 1979. Food and Drugs. Code of Federal Regulations, Title 21, Chapter 178, Part 3620, Subpart B. Washington, D.C.: U.S. Government Printing Office.

Greaves, I.A., R.R. Monson, T.J. Smith, L.J. Pothier, E.A. Eisen, D. Kriebel, S.R. Woskie, S.M. Kennedy, and S. Shalat. 1997a. Respiratory health of automobile workers and exposed to metal-working fluid aerosols. I. Study population, methods, and respiratory symptoms. Am. J. Ind. Med. (submitted).

Greaves, I.A., R.R. Monson, T.J. Smith, L.J. Pothier, E.A. Eisen, D. Kriebel, S.R. Woskie, S.M. Kennedy, and S. Shalat. 1997b. Respiratory health of automobile workers exposed to metal-working fluid aerosols. II. Lung function, conclusions and recommendations. Am. J. Ind. Med. (submitted).

Grose, E.C., M.K. Selgrade, D.W. Davies, and A.G. Stead. 1986. Inhalation Toxicology of Fog Oil Smoke, Final Report. Health Effects Research Laboratory, U.S. Environmental Protection Agency, Research Triangle Park, N.C.

Halder, C.A., T.M. Warne, R.Q. Little, and P.J. Garvin. 1984. Carcinogenicity of petroleum lubricating oil distillates: Effects of solvent refining, hydroprocessing, and blending. Am. J. Ind. Med. 5(4):265-274.

Hendricks, N.V., G.H. Collings, A.E. Dooley, J.T. Garrett, and J.B. Rather, Jr. 1962. A review of exposures to oil mist. Arch. Environ. Health 4:139-145.

Hermann, M., O. Chaude, N. Weill, H. Hofnung, and M. Bedouelle. 1980a. Adaptation of the *Salmonella*/mammalian microsome test to the determination of the mutagenic properties of mineral oils. Mutat. Res. 77:327-339.

Hermann, M., J.P. Durand, J.M. Charpentier, O. Chaude, M. Hofnung, N. Petroff, J.P. Vandecasteele, and N. Weill. 1980b. Correlations of mutagenic activity with polynuclear aromatic hydrocarbon content of various mineral oils. Pp. 899-916 in Polynuclear Aromatic Hydrocarbons: Chemistry and Biological Effects, Fourth International Symposium, A. Bjorseth and A.J. Dennis, eds. Columbus, Ohio: Battelle Press.

Hervin, R.L., and J. Lucas. 1972. Health Hazard Evaluation/Toxicity Determination. Rep. No. HHE-72-35-34. Hazard Evaluation Service Branch, National Institute for Occupational Safety and Health, Rockville, Md. Available from NTIS, Springfield, Va., Doc. No. PB-229-645.

Hodgson, G. 1970. Cutaneous hazards of lubricants. Ind. Med. 39:41-46.

Hodgson, G. 1973. Codes of practice relating to metal working fluids: Health problems arising from contact and exposure of workers to metal working fluids. J. Inst. Petrol. 59:1-8.

Hoekstra, W.G., and P.H. Phillips. 1963. Effects of topically applied mineral oil fractions on the skin of guinea-pigs. J. Invest. Dermatol. 4:79-88.

IARC (International Agency for Research on Cancer). 1984. Mineral oils. Pp. 87-168 in IARC Monographs on the Evaluation of the Carcinogenic Risk of Chemicals to Humans. Polynuclear Aromatic Compounds. Part 2: Carbon Blacks, Mineral Oils, and Some Nitroarenes, Vol. 33. Lyon, France: International Agency for Research on Cancer.

Ingram, A.J., and P. Grasso. 1987. Nuclear enlargement produced in mouse skin by carcinogenic mineral oils. J. Appl. Toxicol. 7:289-295.

Jampolis, R.W., J.R. McDonald, and O.T. Clagett. 1953. Mineral oil granuloma of the lungs: An evaluation of methods for identification

of mineral oil in tissue. Int. Abst. Surg. 97:105-119.
Jarvholm B., L. Lillienberg, G. Sallsten, G. Thiringer, and O. Axelson. 1981. Cancer morbidity among men exposed to oil mist in the metal industry. J. Occup. Med. 23:333-337.
Jarvholm, B., B. Blake, B. Lavenius, G. Thiringer, and R. Vokmann. 1982. Respiratory symptoms and lung function in oil mist-exposed workers. J. Occup. Med. 24:473-479.
Jepsen, J.R., S. Stoyanov, M. Unger, J. Clausen, and H.E. Christensen. 1977. Cutting fluids and their effects on the skin of mice. Acta Pathol. Microbiol. Scand. Sect. A 85:731-738.
Jones, J.G. 1961. An investigation into the effects of exposure to an oil mist on workers in a mill for the cold reduction of steel strip. Ann. Occup. Hyg. 3:264-271.
Kane, M.L., E.N. Ladov, C.E. Holdsworth, and N.K. Weaver. 1984. Toxicological characteristics of refinery streams used to manufacture lubricating oils. Am. J. Ind. Med. 5:183-200.
Katz, S., A. Snelson, R. Butler, R. Farlow, R. Welker, and S. Mainer. 1980. Physical and Chemical Characterization of Military Smokes. Part 2: Fog Oils and Oil Fogs. DAMD17-78-C-8085. AD-A093 205. IIT Research Institute, Chicago, Ill.
Kinnear, J., J. Rogers, O.A. Finn, and A. Mair. 1954. Degenerative changes in the skin with special reference to jute-workers. Br. J. Dermatol. 66:344-349.
Lee, F.K., W.T. Muse, and B.J. Brown. 1989. Mutagenic Responses of Some Petroleum-base Obscurants in the Ames Test. CRDEC-TR-071. Chemical Research, Development and Engineering Center, U.S. Army Armament, Munitions and Chemical Command, Aberdeen Proving Ground, Edgewood, Md.
Liljegren, J.C., W.E. Dunn, G.E. DeVaull, and A.J. Policastro. 1988. Field Measurement and Model Evaluation Program for Assessment of the Environmental Effects of Military Smokes: Field Study of Fog-Oil Smokes. AD-A205 344. Argonne National Laboratory, Argonne, Ill.
Liss-Suter, D., J.E. Villaume, and P.N. Craig. 1978. A Literature Review—Problem Definition Studies on Selected Toxic Chemicals: Occupational Health and Safety Aspects of the Fog Oils SGF No. 1 and SGF No. 2 and Smoke Screens Generated from Them, Vol. 4. AD-A055 903. Science Information Services Department, Franklin Institute Research Laboratories, Rockville, Md.
Lushbaugh, C.C., J.W. Green, and C.E. Redemann. 1950. Effects of

prolonged inhalation of oil fogs on experimental animals. Arch. Ind. Hyg. Occup. Med. 1:237-247.

Mayhew D.A., S.H. Smith, G.L. Doyle, J.C. Kreuger, and K.A. Mellon. 1985. Dermal, Eye and Oral Toxicologic Evaluations of Brass Powder, Fog Oil, Diesel Fuel and Their Mixtures. AD-A172 198. Bioassay Systems Corp., Woburn, Mass.

McKee, R.H., R.T. Plutnick, and R.T. Przygoda. 1989. The carcinogenic initiating and promoting properties of a lightly refined paraffinic oil. Fundam. Appl. Toxicol. 12:748-756.

Okubo, T., and K. Tsuchiya. 1974. An epidemiological study on the cancer mortality in various industries in Japan. Jpn. J. Ind. Health 16:438-452.

Oser, B.L., M. Oser, S. Carson, and S.S. Sternberg. 1965. Toxicologic studies of petrolatum in mice and rats. Toxicol. Appl. Pharmacol. 7:382-401.

Palmer, W.G. 1990. Exposure Standard for Fog Oil. Tech. Rep. 9010. AD-A231 714. U.S. Army Biomedical Research and Development Laboratory, Fort Detrick, Frederick, Md.

Policastro, A.J., and W.E. Dunn. 1985. Survey and Evaluation of Field Data Suitable for Smoke Hazard Model Evaluation. ANL/ER-85-3. AD-A161 880 . Argonne National Laboratory, Ill.

Proudfit, J.P., H.S. van Ordstrand, and C.W. Miller. 1950. Chronic lipid pneumonia following occupational exposure. Arch. Ind. Hyg. Occup. Med. 1:105-111.

Roe, F.J.C., R.L. Carter, and W. Taylor. 1967. Cancer hazard from mineral oil used in the processing of jute. Br. J. Cancer 21:694-702.

Roush, G.C., J.W. Meigs, J.A. Kelly, J.T. Flannery, and H. Burdo. 1980. Sinonasal cancer and occupation: A case-control study. Am. J. Epidemiol. 111:183-193.

Roush, G.C., J.A. Kelly, J.W. Meigs, and J.T. Flannery. 1982. Scrotal carcinoma in Connecticut metalworkers. Sequel to a study of sinonasal cancer. Am. J. Epidemiol. 116:76-85.

Schmal, D., and A. Reiter. 1953. Production of tumors with liquid paraffin, yellow petrolatum and lanolin [in German]. Arzneimittel-Forsch 3:403-406.

Selgrade, M.K., G.E. Hatch, E.C. Grose, J.W. Illing, A.G. Stead, F.J. Miller, J.A. Graham, M.A. Stevens, and J.F. Hardisty. 1987. Pulmonary effects due to short-term exposure to oil fog. J. Toxicol. Environ. Health 21:173-185.

Shoshkes, M., W.G. Banfield, Jr., and S.J. Rosenbaum. 1950. Distribution, effect, and fate of oil aerosol particles retained in the lungs of mice. Arch. Ind. Hyg. Occup. Med. 1:20-35.

Skisak, C.M., C.G. Venier, and D.O. Baker. 1987. Ames tests of lubricating oil products: The mutagenic potency index. In Vitro Toxicol. 1(4):263-276.

Skyberg, K., A. Ronneberg, J.I. Kamoy, K. Dale, and A. Borgerson. 1986. Pulmonary fibrosis in cable plant workers exposed to mist and vapor of petroleum distillates. Environ. Res. 40:261-273.

Sram, R.J., N. Hola, F. Kotesovec, and A. Novakova. 1985. Cytogenetic analysis of peripheral blood lymphocytes in glass workers occupationally exposed to mineral oils. Mutat. Res. 144:277-280.

Stemmer, K.L., and R.W. King. 1979. The Evaluation of the Carcinogenicity of Certain Petroleum Fractions. API Med. Res. Publ. 27-32132. Washington, D.C.: American Petroleum Institute.

Thony, C., J. Thony, M. Lafontaine, and J.C. Limasset. 1975. Concentrations en hydrocarbures polycycliques aromatiques cancerogenes de quelques huiles minérales. Étude du resque correspondent. Arch. Mal. Prof. Med. Travail Secur. Soc. 36:37-52.

Thony, C., J. Thony, M. Lafontaine, and J.C. Limasset. 1976. Hydocarbures polycycliques aromatiques cancerogenes dans les produits petroliers preventions possibles du cancer des huiles minerales. INSERM Symposia Series. IARC Sci. Publ. 52(13):165-170.

Vena, J.E., H.A. Sultz, R.C. Fiedler, and R.E. Barnes. 1985. Mortality of workers in an automobile engine and parts manufacturing complex. Br. J. Ind. Med. 42(2):85-93.

Venier, C.G., C.M. Skisak, and D.A. Bell. 1987. Ames tests of lubricating oil products: The effect of processing variables. In Vitro Toxicol. 1(4):253-261.

Wagner, W.D., P.G. Wright, and H.E. Stokinger. 1964. Inhalation toxicology of oil mists. I. Chronic effects of white mineral oil. Am. Ind. Hyg. Assoc. J. 25:158-168.

Waterhouse, J.A.H. 1971. Cutting oils and cancer. Ann. Occup. Hyg. 14(2):161-170.

Young, J.Y., D.A. Smart, J.T. Allen, D.L. Parmer, A.B. Rosencrance, E.E. Brueggeman, and F.H. Broski. 1989. Field Exposure of Chemical School Students and Cadre to Fog Oil and Hexachloroethane (HC) Smokes. Tech. Rep. 8908. U.S. Army Biomedical Research and Development Laboratory, Fort Detrick, Frederick, Md.

4

Red Phosphorus Smoke

BACKGROUND INFORMATION

The military application of phosphorus smokes for environmental screening can contain either white phosphorus or red phosphorus in various matrices (e.g., felt, butyl rubber, or polymer epoxy binders). The compositions of the various phosphorus smokes are similar, being composed primarily of polyphosphoric acids with less than 1% trace levels of organic compounds. This chapter provides an evaluation of the health effects of red phosphorus (RP) in combination with butyl rubber (BR).

Military Applications

Red phosphorus is not manufactured by the U.S. Army and is obtained from sources outside the United States (e.g., People's Republic of China) and shipped to Pine Bluff Arsenal in Arkansas for blending and filling operations for certain munitions and testing (Mitchell and Burrows 1990). In Army field use, red phosphorus smoke is deployed explosively from grenades and mortar shells.

The obscurant portion of the grenades consists of a 95:5 mixture of red phosphorus and styrene-butadiene rubber (butyl rubber) in the presence of methylene chloride, which is later removed by low temperature drying (Lundy and Eaton 1994). Analysis of samples for methylene chloride found none present (Brazell et al. 1984). The purpose of the butyl rubber is to reduce the cloud-pillar effect found with pure red phosphorus. This mixture of red phosphorus and butyl rubber (RP-BR) also contains two other compounds. The red phosphorus is coated with about 1.25% (by weight) of insulating oil, and approximately 1% talc or silica is added to break up and improve uniformity of the pattern.

Red phosphorus is also the major ingredient in mortar rounds used to generate smoke. In that use, it is combined with sodium nitrate and an epoxy binder in a ratio of 80:14:6 parts by weight, respectively.

Physical and Chemical Properties

The red allotropic form of elemental phosphorus is intermediate to the black and white varieties in reactivity. Red phosphorus reacts slowly with oxygen and water vapor and can evolve phosphine gas, which is highly toxic. The reaction is extremely slow at normal temperatures and humidities and is not considered to be a factor in the deployment of phosphorus munitions in military operations. However, this reaction can be catalyzed by metal ions (e.g., iron and copper), which can markedly increase the oxidation rate. The physical and chemical properties of red phosphorus are listed below:

Formula:	Polymeric $(P_4)_n$
CAS no.:	7723-14-0
Molecular weight:	123.9_n
Density:	2.34 g/cm^3
Melting point:	Sublimes at 416°C

Heat of sublimation:	19.7 kcal/mol
Ignition temperature:	280°C in air
Solubility:	Insoluble in organic solvents, negligible in water

Occurrence and Use

Phosphorus is the 11th most abundant element in the earth's crust. It commonly occurs in the lithosphere in igneous, sedimentary, and metamorphic rock. Depending upon the nature of interatomic bonds established during its formation, solid elemental phosphorus can occur in three allotropic forms: black, white, or red.

In the biosphere, phosphorus is an essential nutrient in the formation of structural biomolecules, such as membrane phospholipids; functional macromolecules, such as nucleic acids and adenosine triphosphate; and metabolic intermediates, such as sugar phosphates. Human activities, such as the manufacture and use of detergents, fertilizers, and water softeners, contribute to phosphorus in the environment.

Combustion Products

The combustion products associated with RP-BR have been chemically and physically characterized by the U.S. Army. Table 4-1 summarizes the composition of phosphoric acids in RP-BR smoke. The combustion products of RP-BR and white phosphorus impregnated in felt, also used by the Army to generate an obscuring smoke, are similar under the same burn conditions (Ramsey et al. 1985). The particles are composed primarily of various phosphoric acids present as a complex mixture of polymeric forms with organic compounds and inorganic gases only at trace levels. Phosphorus trioxide is particularly likely to be formed, which is of interest because it reacts with water to form phosphoric acid and phosphine (Ballou 1981). The relative proportion of the different acids of phosphorus in RP-BR smoke changes with time after it is generated, but the predominate component of the smoke remains phosphoric acid (orthophosphate) (Ballou 1981; Mitchell and Burrows 1990). Only trace amounts of phosphine have been measured in some cases (Ballou 1981).

TABLE 4-1 Composition of the Phosphoric Acids in RP-BR Smoke from Static Burn

Component	Composition %
Orthophosphate	22.8
Pyrophosphate	19.6
Tripolyphosphate	13.3
Tetrapolyphosphate	11.5
P_5-P_{13}	32.8
Higher polyphosphates	Low

Source: Brazell et al. (1984).

Measurement of the mass median aerodynamic diameter (MMAD) of RP-BR aerosol particles generated to test the toxicity of the smoke in animals have ranged from 0.4 to 1.6 μm, and geometric standard deviations ranged from 1.5 to 1.9 (Aranyi 1984, Aranyi et al. 1988b; Ballou 1981). All measurements of RP-BR smoke reported or recommended in this chapter are referred to in milligrams of total particulates per cubic meter.

TOXICOKINETICS

No studies were available on the toxicokinetics of RP-BR smoke or its components.

TOXICITY SUMMARY: ELEMENTAL RED PHOSPHORUS

Red phosphorus is a relatively inactive allotrope of phosphorus. Little, if any, significant toxicity appears to be associated with elemental red phosphorus unless it is contaminated with white phosphorus (Mitchel and Burrows 1990; Lundy and Eaton 1994). No dermal irritation was noted when red phosphorus was applied to the skin of rabbits at doses of 0.5 g per site. Similarly, dermal application of the element to guinea pigs resulted in no skin irritation or sensitization. Interdermal injection resulted in only slight irritation. Doses of 100 mg did not result in rabbit eye irritation. In Fischer 344 rats, the oral LD_{50} was reported to be greater than 10 g/kg (Mitchell and Burrows 1990).

TOXICITY SUMMARY: RED PHOSPHORUS-BUTYL RUBBER

Effects in Humans

No studies have been conducted on the effects of RP-BR smoke in humans. However, Mitchell and Burrows (1990) estimated that human exposure to RP-BR at concentrations of about 2,000 mg/m^3 for longer than 15 min might result in death. They suggested further that acute exposure at concentrations of 1,000 mg/m^3 should be considered intolerable and that 700 mg/m^3 is the highest tolerable concentration; above that, masks must be worn (Mitchell and Burrows 1990). Others have reported that concentrations exceeding 100 mg/m^3 were unendurable for all workers except the "inured" worker (ACGIH 1991).

Effects in Animals

Dermal Exposures

When air samples of combusted RP-BR were collected using an electrostatic precipitator and the residue (0.1 mL) was instilled in the eyes of rabbits, severe irritation and corneal ulceration were evident. When this same material was administered to the intact or abraded skin of rabbits clipped free of hair, severe irritation was produced (Weimer et al. 1977).

Inhalation Exposures

One-Time Exposures

Lethality. The concentration lethal to 50% (LC_{50}) of Sprague-Dawley rats that were exposed to RP-BR smoke for 1 hr on 5 consecutive days was estimated to be 2,320 mg/m^3 (Aranyi et al. 1988a). In contrast, daily 4-hr exposures for 5 days at lower concentrations (i.e., W 1,000 mg/m^3) did not produce significant mortality (Aranyi et al. 1988a). Other reports estimated that a single 1-hr LC_{50} for rats was approximately 4,000 mg/m^3 (Ballou 1981; Shinn et al. 1985). The presence of butyl rubber in the smoke was assumed to be toxicologically insignificant. In both reports, the animals continued to die during the 14-day observation period following exposure, indicating that both acute and delayed effects resulted from these 1-hr exposures. Twenty percent died after a 1-hr exposure at 3,100 mg/m^3, and eight of nine died within 2 days of a 1-hr exposure at 8,460 mg/m^3 (Ballou 1981). Lethal effects were accompanied by symptoms of respiratory distress, including pulmonary edema, congestion, and atelectasis. Ballou (1981) found that the application of Haber's law (i.e., effects are proportional to the exposure concentration (C) multiplied by the duration (or time, T) of the exposure, or the CT product) predicted lethality of such smokes with a 65% error. Several studies indicate that exposure concentration, instead of the CT product, is the major determinant of lethality over relatively short exposures (Shinn et al. 1985; Aranyi et al. 1988a; Lundy and Eaton 1994).

Weimer et al. (1977) studied the acute effects of a single exposure to RP-BR smoke with a wide range of exposure CT values in three species of animals. For rats, average exposure concentrations ranged from 1,128 to 1,882 mg/m^3, and exposure durations ranged from 60 to 240 min, producing CT values of 67,685 to 451,680 mg·min/m^3. For guinea pigs, exposure concentrations ranged from 120 to 2,277 mg/m^3, and exposure durations ranged from 5 to 150 min, producing CTs from 45,570 to 451,680 mg·min/m^3. For dogs, exposure concentrations ranged from 1,212 to 1,882 mg/m^3, and exposure durations ranged from 30 to 240 min, producing CTs from 45,570 to 451,680 mg·min/m^3.

Weimer et al. (1977) found the CT products lethal to 50% (LCT_{50}) of rats, guinea pigs, and dogs tested were 222,700, 4,040, and more than 451,700 mg·min/m^3, respectively. The exposure time in these studies varied from 5 min to 240 min. No rats died following exposure at 1,128 mg/m^3 for 60 min, and 4 of 10 died after exposure at 1,676 mg/m^3 for 120 min. For the guinea pig, 4 of 10 died from a 10-min exposure at 352 mg/m^3; whereas none died from a 5-min exposure at 120 mg/m^3. No deaths occurred among dogs exposed to the highest dose studied (1,882 mg/ for 240 min). These studies indicate that RP-BR smoke is only slightly toxic to the rat and dog, but that the guinea pig is more sensitive. Table 4-2 summarizes the acute lethality data.

Skin and Eye Irritation. Weimer et al. (1977) reported conjunctivitis in rats exposed at 1,813 mg/m^3 for 180 min and in dogs exposed at 1,882 mg/m^3 for 240 min. No conjunctivitis was apparent 3 days after exposure ended. In the rat, however, that exposure concentration resulted in 9 of 10 deaths over a 6-day period after exposure (Weimer et al. 1977).

Pulmonary Effects. Ballou (1981) found that gross pathology in rats following exposure to high concentrations of airborne RP-BR smoke varied somewhat with concentration, but the pathology consistently involved the laryngeal and proximal tracheal regions. As indicated in Table 4-2, at the highest exposure concentration tested (8,460 mg/m^3 for 1 hr), eight of nine animals died within 2 days. Seven of those had significant laryngeal and tracheal lesions that consisted of a fine fibrinlike coat on the laryngeal and proximal tracheal mucosa. Pulmonary edema and hemorrhage were prominent. The only rat that survived the exposure had laryngeal edema, small fibrin tags in the central larynx, and essentially no epiglottis. All the other concentrations tested (5,360, 4,330, and 3,150 mg/m^3 for 1 hr and 1,530 mg/m^3 for 4 hr) also resulted in some lethality in the exposed groups. At all these concentrations, the animals showed marked laryngeal and epiglottal erosion, hemorrhage, ulceration with fibrin deposition, enlarged lymph nodes, edema with pulmonary congestion, and mucus accumulation in the trachea.

TABLE 4-2 Acute Lethality of Red Phosphorus-Butyl Rubber Smoke

Species	Exposure Duration	Exposure Concentration (mg/m^3)	End Points and Comments	Reference
Rat	1 hr/d, 5 d	2,320	LC_{50}	Aranyi et al. 1988a
Rat	4 hr/d, 5 d	W 1,000	No significant mortality	Aranyi et al. 1988a
Rat	1 hr	4,000	LC_{50}	Ballou 1981; Shinn et al. 1985
Rat	1 hr	8,460	8/9 died with 2 d	Ballou 1981
Rat	1 hr	3,100	20% lethality	Ballou 1981
Rat	60 min	1,128	No deaths	Weimer et al. 1977
	120 min	1,676	4/10 died	
	150 min	1,625	5/10 died	
	180 min	1,572	8/10 died	
Guinea pig	5 min	120	No deaths	Weimer et al. 1977
	10 min	352	4/10 died	
	15 min	484	7/10 died	
	10 min	797	9/10 died	
Dog	W 240 min	W 1,882	No deaths	Weimer et al. 1977

In the series of experiments conducted to estimate LCT_{50} values for rats, guinea pigs, and dogs, Weimer et al. (1977) stated that the animals were not visible for observation during exposure. Following exposure, however, all animals displayed signs of respiratory distress at all exposure concentrations. The lowest exposure concentrations tested for the rat, guinea pig, and dog were 1,128 mg/m^3 for 60 min, 120 mg/m^3 for 5 min, and 1,519 mg/m^3 for 30 min, respectively. Animals were hypoactive and salivating. As the CT values increased, distress became more marked. Those effects persisted for up to 2 days after exposure.

Following a one-time 3.5-hr exposure of rats at 1,000 mg/m^3, Aranyi et al. (1988b) observed a large reduction in pulmonary bacteriocidal activity, from 80% activity in the controls to 35% activity in the exposed group.

Other Effects. Weimer et al. (1977) also reported some extrapulmonary effects following inhalation of RP-BR. Male rats exposed at 1,676 mg/m^3 for 120 min had kidney weights that were significantly less than control weights. Also, body weights were significantly lower after exposure at 1,625 mg/m^3 for 150 min. Both of those concentrations also produced significant mortality. The authors stated that the gross pathology and histopathological evaluations of the rats, guinea pigs, and dogs failed to show any lesions that could be attributed to the smoke inhalation. No significant changes in blood hematology or chemistry developed in dogs, guinea pigs, or rats that could be agent-related.

Repeated Exposures

Skin and Eye Irritation. Weimer et al. (1980) reported transient ocular irritation in rats exposed to RP-BR smoke concentrations of 0 (control), approximately 22 mg/m^3, and approximately 165 mg/m^3 for 8 min per day, 5 days per week, for 12 weeks (60 rats per group). During the eighth week, a reddening and swelling of the eyelids was noted in rats at both concentrations. Those effects subsided by the end of the exposure. Several control rats also displayed similar reddened eyelids. The number of animals with eye irritation exposed to the control, low, and high concentrations of RP-BR smoke was 3, 8, and 14, respectively. Although no statistical analyses were performed, the dose-response relation suggests that this effect was attributable to the RP-BR exposure.

Pulmonary Effects. Weimer et al. (1980) also examined the pulmonary effects of repeated exposures to two concentrations of RP-BR. Two strains of rats (Sprague-Dawley and Fischer 344), two strains of mice (Swiss and A-strain), guinea pigs, and rabbits were used. The test animals were exposed for 5 days per week over 12 weeks. The low exposure concentration resulted in a cumulative CT of 10,705 mg•min/m^3 and a mean daily exposure CT of 178 mg•min/m^3. The mean daily exposure time was about 8 min. At the high exposure concentration, the cumulative CT was 81,691 mg•min/m^3; the mean daily exposure CT was 1,319 mg•min/m^3, and the mean daily exposure duration was 8 min. The daily exposure concentrations ranged from 8 to 43 mg/m^3 (average 22 mg/m^3) for the low exposure concentration. For the high exposure concentration, RP-BR concentrations were between 80 and 288 mg/m^3 daily (average 165 mg/m^3).

During the first 3 days of exposure, both strains of rats displayed an increase in breathing rates following exposure to either concentration. Although histological changes were observed in the lungs, trachea, upper respiratory tract, and other organs in both rats and mice, Weimer et al. (1980) stated that these changes were not unlike those observed in the controls. Moreover, the changes were sporadic, and the incidence or severity of the changes was not related to the exposure concentration. The authors concluded that the pathological changes identified could not be attributed to exposure to the smoke. Some of the animals were held for 24 months in clean air after the exposure. No evidence of latent toxic effect or exposure-related tumor formation was found that could be agent-related. However, no long-term cancer bioassays have been conducted on RP-BR smoke aerosol.

At both concentrations, the rabbits and guinea pigs exposed

to RP-BR had a number of morphological lesions in the lung, trachea, nasal turbinates, liver, kidney, heart, testes, ovaries, urinary bladder, and other organs, but these changes also were seen in the controls. Weimer et al. (1980) stated that these changes could not be attributed to the test substance. The high incidence of pathology in the control animals might have resulted from one or more of several factors, including disease, age or source of the animals, or animal housing conditions. Pulmonary function tests performed on exposed guinea pigs indicated that after 3 weeks of exposure, pulmonary resistance decreased at the high and low exposure concentrations, but only in male guinea pigs. This effect was not present following 6, 9, or 12 weeks of exposure. Weimer et al. (1980) concluded that animals exposed repeatedly to RP-BR did not experience short-term or cumulative toxic effects. Thus, the high exposure concentration, 165 mg/m^3, could be considered a no-observed-adverse-effect level (NOAEL) for all strains and species of animals tested.

In another set of studies, male and female Sprague-Dawley rats were exposed to RP-BR aerosols ranging in concentrations from 400 to 1,200 mg/m^3 for 2.25 hr per day, 4 days per week, for 4 weeks (Aranyi 1983, 1984; Lundy and Eaton 1994). During the exposure, wheezing and labored breathing were observed in the male rats exposed at the high dose. Decreased body weights and reduced food consumption were seen in the male rats during the 4-week exposure, but those conditions returned to normal during the 14-day recovery period.

These authors also examined pulmonary free cells collected by lung lavage and found a slight, but not statistically significant, increase in total number of free cells immediately following exposures at 750 mg/m^3 (Aranyi 1983, 1984; Aranyi et al. 1988b). After a 14-day recovery period, the count returned to normal. A significant increase in the protein level in the pulmonary lavage fluid of rats of both sexes after exposure to more than 1,000 mg/m^3 indicated pulmonary edema, which was resolved during the recovery period (Aranyi 1984).

The primary lesion of the respiratory tract was terminal

bronchiolar fibrosis, which was evident after exposure at 400 mg/m^3 for 3.5 hr per day for 4 consecutive days (Aranyi 1983). The lesion increased in incidence and severity with increasing exposure concentrations and duration, and animals did not exhibit any recovery during the 14 days in clean air. The thickening of the terminal bronchioles and associated alveolar walls was due to the formation of new collagen fibers. Peribronchiolar and perivascular infiltration of eosinophils occurred but regressed during the recovery period (Aranyi 1983). In a separate study, rats exposed at 750 or 1,000 mg/m^3 for 2.5 hr per day, 4 consecutive days per week, for 4 weeks resulted in minimal-to-mild terminal bronchiolar fibrosis, but no effect was seen in rats exposed at 400 mg/m^3 (Aranyi 1984).

On the basis of these data, investigators conducted two additional studies using the male rat only. The length of exposure was 13 weeks, and the concentrations tested included 50, 180, 300, 750, and 1,200 mg/m^3. The exposures were for 2.25 hr per day for 4 days per week (Aranyi 1986; Aranyi et al. 1988a; Lundy and Eaton 1994). Statistically significant decreases in body weights were observed from weeks 1 through 13 in the groups exposed at 750 and 1,200 mg/m^3. Food consumption decreased significantly. Of the animals exposed at 1,200 mg/m^3, 10.8% died spontaneously or were necropsied in a moribund state. Most of the animals died during the first 2 weeks of exposure and had varying degrees of congestion and small amounts of hemorrhage in the lungs. Animals exposed at 750 and 1,200 mg/m^3 that died later in the study had terminal bronchiolar fibrosis and erosions of the laryngeal mucosa with deposition of fibrin on the surface. No exposure-related deaths occurred in the groups exposed at less than 300 mg/m^3 (Aranyi 1986).

Any significant changes in pulmonary lavage fluid found after either 4 or 13 weeks of exposure were absent after 8 weeks of recovery in clean air (Aranyi 1983, 1984, 1986; Aranyi et al. 1988a; Lundy and Eaton 1994), indicating that the macrophages returned to their normal state within 8 weeks of exposure. A significant decrease in pulmonary bactericidal activity seen at concentrations of 300 mg/m^3 and above were also completely absent after the recovery period (Aranyi et al. 1988b).

Histologically, the primary exposure-related change seen after termination of the studies was the presence of terminal bronchiolar fibrosis, resulting in a thickening of the alveolar walls and of the most distal portions of the terminal bronchioles at the site where they join the alveolar sacs. Microscopic examination of the lungs showed that at 2 weeks of exposure, 50% of the rats exposed at 750 mg/m^3 had minimal fibrosis, and all of the rats exposed at 1,200 mg/m^3 had minimal to mild fibrosis. All rats had fibrosis after 4 weeks of exposure at 750 and 1,200 mg/m^3 . After the completion of the 13-week study, 100% of the rats exposed at 750 mg/m^3 and higher and approximately 50% of the rats exposed at 300 mg/m^3 had terminal bronchiolar fibrosis. Minimum fibrosis was found in approximately 25% of the rats exposed at 180 mg/m^3. At 50 mg/m^3, the lowest concentration tested, no changes were found. The NOAEL for terminal bronchiolar fibrosis from exposure to RP-BR was 50 mg/m^3, and the lowest-observed-adverse-effect level (LOAEL) was 180 mg/m^3 (Aranyi 1986; Aranyi et al. 1988a).

Biochemical Effects. In the studies of Sprague-Dawley rats exposed to RP-BR aerosols at concentrations ranging from 400 to 1,200 mg/m^3 , 2.25 hr per day, 4 days per week, for 4 weeks (Aranyi 1983, 1984), decreased cholesterol and blood urea nitrogen (BUN) values were seen in all RP-BR-exposed males (750 mg/m^3 was the lowest concentration tested). In addition, concentration-related decreases in BUN, cholesterol, and triglycerides levels were seen in all RP-BR-exposed females immediately after exposure (400 mg/m^3 was the lowest concentration tested). After the recovery period, only female rats exposed at concentrations of 1,000 mg/m^3 showed significantly decreased cholesterol and triglyceride levels, and female rats exposed at more than 750 mg/m^3 showed decreased BUN levels.

Immunological Effects. In the same studies using the 4-week exposure period (Aranyi 1983, 1984; Lundy and Eaton 1994), at concentrations of 750 mg/m^3, the white-blood-cell (WBC) counts decreased in male rats by the end of the exposure. Increased blood lymphocytes were also seen at the same concentration in the female rats during and after the recovery period. No treatment-related histopathological changes were found outside the respiratory tract (Aranyi 1983, 1984; Lundy and Eaton 1994).

Significant increases in adenosine 5'-triphosphate (ATP) levels in macrophages lavaged from the lung, expressed as ATP/105 cells or ATP/total protein, were observed immediately after the last exposure at 750, 1,000, and 1,200 mg/m^3 for male rats and at 400 and 750 mg/m^3 for female rats in the 4-week exposure experiments (Aranyi 1984). After recovery, ATP/total protein from male rats exposed at the high dose remained increased, whereas ATP/10^5 cells and ATP/total protein were increased in macrophages from female rats exposed at 1,000 mg/m^3 . A consistent finding was decreased activity of the plasma membrane-associated ectoenzyme 5'-nucleotidase in macrophages in rats exposed at a concentration of at least 750 mg/m^3 (Aranyi 1984; Aranyi et al. 1988b). In addition, macrophages of male rats tested after the 14-day recovery also had decreased alkaline phosphatase activity. Decreased activity of 5'-nucleotidase and alkaline phosphatase in macrophages has been associated with enhanced in vitro antitumor and antiviral activity (Lundy and Eaton 1994). These data suggest that a change in alveolar macrophage populations might be induced by exposure to RP-BR that activated these cells (Aranyi et al. 1988a). ATP levels and 5'-nucleotidase activity had returned to normal when the recovery period was extended to 8 weeks. In the 13-week exposure paradigm, increased cellular ATP levels occurred at the lowest exposure concentration tested (300 mg/m^3), but decreased activity of 5'-nucleotidase occurred only at exposure concentrations of 750 mg/m^3 or higher (Aranyi et al. 1988b).

Neurobehavioral Effects. Of the neurobehavioral variables studied, only locomotor activity was significantly affected by the exposure. Male rats showed increased motor activity at all concentrations and incomplete recovery after 2 weeks at some concentrations. Females showed a trend toward increased activity but no evidence of such effect after the recovery period (Aranyi 1983, 1984).

Reproductive and Developmental Effects. Weimer et al. (1980) exposed Sprague-Dawley rats 5 days per week for 10 weeks to RP-BR at concentrations of 132 or 1,186 mg•min/m^3 and observed no dominant lethal or single-generation reproductive effects. The mean daily exposure duration was approximately 8 min (Weimer et al. 1980; Lundy and Eaton 1994). Weimer et al. (1980) also exposed pregnant rats 5 days per week from gestation days 6 through 15 to RP-BR smoke at 132 or 1,186 mg•min/m^3. The mean daily exposure duration was 8 min. The fetuses were examined for skeletal and visceral anomalies. No dose-related increases were seen in any malformation or variations. In a single-generation study using the same exposure concentrations, offspring body weights were decreased on postnatal day (PND) 1 in the high-dose group, with a rebound in body weights in these pups at PNDs 14 and 21. Low-dose male and female offspring were heavier than controls at PNDs 4 to 21 (Weimer et al. 1980). No information was provided on body-weight gain or fertility in adults or on viability, survival, or lactation indices for the above studies. In a separate 12-week exposure study, nonpregnant Sprague-Dawley females in the high-dose group exhibited a significantly lower weight gain after 4 weeks of exposure than females in the control or low-dose groups (Weimer et al. 1980). No effects were reported on testicular toxicity in any of these studies. However, it is not clear what fixative was used to judge histopathology, and formalin fixative is inadequate for testicular tissue. Because of the sparse data, a NOAEL cannot be established in terms of possible male and female reproductive toxicity.

Carcinogenic and Mutagenic Effects. No long-term carcinogenicity studies for RP-BR smokes have been conducted. Micronucleus analysis was performed on bone-marrow polychromatic and normachromatic red blood cells (RBC) and on circulating RBC of female rats exposed 8 times over a 2-week period to RP-BR at 1,000 mg/m^3 for 2.25 hr (Aranyi 1984). The conclusion of the author was that RP-BR is a weak clastogen in the micronucleus test. The results showed a significant clastogenic response in both bone marrow and RBC of rats that were exposed for 2 weeks, but that effect was not found after a 4-week exposure or after a 2-week recovery period. Effects after the 4-week exposure and 2-week recovery period would not be expected because micronuclei are removed from the circulation after 24 to 30 days. The fact that micronuclei are not observed after a 4-week exposure or a 2-week recovery period does not diminish the significance of observations at the end of the 2-week exposure period. These results are consistent with the conclusion that RP-BR is a weak clastogen. These results alone, however, do not allow a conclusion that RP-BR is mutagenic.

Summary of Toxicity Data

Table 4-2 (above) and Table 4-3 (below) summarize the lethal and nonlethal effects of exposure to RP-BR aerosols.

Noncancer Toxicity

Phosphorus smoke aerosols act as irritants because of their high phosphoric acid content. Respiratory irritation and inflammation have been noted in humans and in animal studies.

The effect occurring at the lowest short-term exposure concentration is respiratory distress. Symptoms including labored breathing, hypoactivity, salivation, and redness of the eyes have been reported at exposure concentrations of 1,128 mg/m^3 for 1 hr for rats, 1,519 mg/m^3 for 30 min for dogs, 1,212 mg/m^3 for 90 min for dogs, and 120 mg/m^3 for 5 min for guinea pigs (Table 4-3). Reports have indicated that human exposure to concentrations ranging from 100 to 1,000 mg/m^3 can be intolerable.

TABLE 4-3 Summary of Nonlethal Effects of Exposure to Red Phosphorus-Butyl Rubber Smoke Aerosol

Category and Species	Exposure Frequency and Duration	NOAEL (mg/m^3)	LOAEL (mg/m^3)	End Point and Comments	Reference
Effects in Humans					
Human (workers)	8 hr/d, 5 d/wk, several years	—	100	Unendurable except for the "hardened" worker	ACGIH 1991
Effects in Animals					
One-Time Inhalation Exposures					
Skin and Eye Irritation					
Rat	180 min	—	1,813	Conjunctivitis	Weimer et al. 1977
Dog	240 min	—	1,882	Conjunctivitis	Weimer et al. 1977
Pulmonary Effects					
Rat	1 hr	—	3,150	Laryngeal and tracheal lesions; pulmonary edema (and some lethality)	Ballou 1981
Rat	4 hr	—	1,530	Laryngeal and tracheal lesions; pulmonary edema (and some lethality)	Ballou 1981
Rat	1 hr	—	1,128	Respiratory distress; hypoactivity; salivation	Weimer et al. 1977

Rat	3.5 hr, one time	–	1,000	Decrease from 80% (control) to 35% bactericidal activity; reversible in clean air (only one exposure concentration tested)	Aranyi et al. 1988b
Guinea pig	5 min	–	120	Respiratory distress; hypoactivity; salivation	Weimer et al. 1977
Dog	30 min	–	1,519	Respiratory distress; hypoactivity; salivation	Weimer et al. 1977
	90 min	–	1,212		
Other Effects					
Rat	120 min	–	1,676	Reduced kidney weight (and some lethality)	Weimer et al. 1977
Rat	150 min	–	1,625	Reduced body weight (and some lethality)	Weimer et al. 1977
Repeated Inhalation and Ocular Exposures					
Eye Irritation					
Rat	8 min/d, 5 d/wk, 12 wk	–	22	Transient reddening and swelling of eyelids during 8th wk only	Weimer et al. 1980
Pulmonary Effects					
Rat (Sprague-Dawley and Fischer)	8 min/d, 5 d/wk, 12 wk	165	–	No effects	Weimer et al. 1980
Mice (Swiss and (A strain)	8 min/d, 5 d/wk, 12 wk	165	–	No effects	Weimer et al. 1980
Guinea pig	8 min/d, 5 d/wk, 12 wk	165	–	No effects	Weimer et al. 1980

Category and Species	Exposure Frequency and Duration	NOAEL (mg/m^3)	LOAEL (mg/m^3)	End Point and Comments	Reference
Rabbit	8 min/d, 5 d/wk, 12 wk	165	–	No effects	Weimer et al. 1980
Rat	3.5 hr, 4 d	–	400	Terminal bronchiolar fibrosis	Aranyi 1983
Rat	2.25 hr/d, 4 d/wk, 4 wk	C : 750 X : 1,000	C : 1,000 X : 1,200	Increased protein in lavage fluid	Aranyi 1984
Rat	2.25 hr/d, 4 d/wk, 4 wk	400	750	Terminal bronchiolar fibrosis	Aranyi 1984
Rat	2.25 hr/d, 4 d/wk, 13 wk	–	300	Decreased total cells in pulmonary lavage fluid; reversible in clean air	Aranyi et al. 1988b
Rat	2.25 hr/d, 4 d/wk, 13 wk	180	300	Reduced bactericidal activity; reversible in clean air	Aranyi et al. 1988a,b
Rat	2.25 hr/d, 4 d/wk, 13 wk	50	180	Terminal bronchiolar fibrosis	Aranyi 1986; Aranyi et al. 1988a
Biochemical Effects					
Rat	2.25 hr/d, 4 d/wk, 4 wk	–	C: 400 X: 750	Decrease in cholesterol and BUN levels	Aranyi 1984
Immunological Effects					
Rat	2.25 hr/d, 4 d/wk, 4 wk	400	750	Decreased white-blood-cell count	Aranyi 1984

Rat	2.25 hr/d, 4 d/wk, 4 wk	–	C: 400 X: 750	Increased cellular ATP levels	Aranyi 1984
Rat	2.25 hr/d, 4 d/wk, 4 wk	–	750	Decreased activity of 5'-nucleotidase	Aranyi 1984; Aranyi et al. 1988b
Rat	2.25 hr/d, 4 d/wk, 13 wk	–	300	Increased cellular ATP levels	Aranyi et al. 1988b
Rat	2.25 hr/d, 4 d/wk, 13 wk	300	750	Decreased activity of 5'-nucleotidase	Aranyi et al. 1988b
Behavioral Effects					
Rat	2.25 hr/d, 5 d/wk, 4 wk	–	400	Increased motor activity; incomplete recovery after 2 wk in clean air	Aranyi 1983, 1984
Reproductive and Developmental Effects					
Rat	8 min/d, 5 d/ wk, 10 wk	132	1,186	Decreased birth weight (reproductive end points not fully evaluated)	Weimer et al. 1980; Lundy and Eaton 1994
Mutagenic Effects					
Rat	2.25 hr/d, 4 d/wk, 2 wk	–	1,000	Clastogenic response	Aranyi 1984

Abbreviations: hr, hour(s); min, minute(s); d, day(s); wk, week(s); X, male; C, female.
Notes: Aranyi (1984) used exposure concentrations of 400, 750, and 1,000 mg/m^3 for females and 750, 1,000, and 1,200 mg/m^3 for males. Thus, if an effect was observed at all concentrations tested, the LOAEL (without a NOAEL) would be 400 mg/m^3 for females and 750 mg/m^3 for males. Aranyi (1983, 1984) and Aranyi et al. (1988a,b) found no differences in responses between male and female rats exposed for 2.25 hr/d, 4 d/wk for 4 wk; therefore, only male rats were used in the 13-wk exposure experiments.

A major health concern for repeated inhalation exposures to RP-BR smoke aerosol is development of terminal bronchiolar fibrosis. That condition is irreversible. The induction of fibrosis appears to be influenced by both concentration and duration of exposure. After 2 weeks of exposure, 50% of male rats exposed at 750 mg/m^3 had minimal fibrosis, and all the test animals exposed at 1,200 mg/m^3 had minimal-to-mild fibrosis. After 4 weeks of exposure, all rats exposed at 750 and 1,200 mg/m^3 exhibited fibrosis. After 13 weeks of exposure, 100% of the rats exposed at 750 mg/m^3 or more and 30% of the rats exposed at 300 mg/m^3 had terminal bronchiolar fibrosis, and 100% of the rats exposed at 180 mg/m^3 exhibited minimal fibrosis. At 50 mg/m^3, the lowest concentration tested, no changes were found. Based on these studies, the NOAEL for terminal bronchiolar fibrosis in rats was 50 mg/ and the LOAEL was 180 mg/m^3.

Carcinogenicity

There is no evidence of carcinogenicity or mutagenicity of RP-BR smoke; however, few tests have been conducted to examine these end points.

EXISTING RECOMMENDED EXPOSURE LIMITS

The American Conference of Governmental Industrial Hygienists (ACGIH 1991) Threshold Limit Values (TLVs), both time-weighted-average (TWA) values (for 8 hr per day, 5 days per week, for 40 years) and short-term exposure limits (STELs), for various components of RP-BR smokes are listed in Table 4-4. ACGIH (1991) recommended the TLV-TWA for phosphoric acid by analogy to comparable experience and data for sulfuric acid. ACGIH (1991) observed that the TLV-TWA is below the concentration that causes throat irritation among unacclimated workers and is well below the concentration that is well tolerated by acclimated workers.

TABLE 4-4 ACGIH Recommended Exposure Limits

RP-BR Smoke Component	TLV-TWA	TLV-STEL
Phosphine	0.3 ppm (0.42 mg/m^3)	1.0 ppm (1.4 mg/m^3)
Phosphoric acid	1.0 mg/m^3	3.0 mg/m^3

Abbreviations: ACGIH, American Conference of Governmental Industrial Hygienists; TLV, Threshold Limit Value; TWA, time-weighted average; STEL, short-term exposure limit.

SUBCOMMITTEE EVALUATION AND RECOMMENDATIONS

Using the toxicity information described above, the subcommittee developed exposure guidance levels for military personnel exposed during an emergency release and during regular training exercises and for consideration at military-training-facility boundaries to protect nearby communities from emergency or repeated releases of RP-BR smoke.

Military Exposures

Emergency Exposure Guidance Level (EEGL)1

The EEGLs are based on the subcommittee's interpretation of available information in the context of an emergency, when some risk of reversible health effects and discomfort are considered acceptable.

[1] Guidance for a rare, emergency situation resulting in exposure of military personnel.

The major toxic end points associated with such short-term exposures would be lethality, respiratory distress and irritation, and pulmonary lesions. All the other effects documented for RP-BR smoke were from studies using longer-term exposures, and extrapolation from those to a 15-min, 1-hr, or 6-hr EEGL would be inappropriate.

The most sensitive response to short-term exposures to RP-BR aerosols is respiratory distress. Data from humans indicate that concentrations as low as 100 mg/m^3 can be considered intolerable (Table 4-3). Given the lack of documentation by ACGIH (1991) on how the human LOAEL of 100 mg/m^3 was determined, the subcommittee used the animal data to estimate EEGLs. Because the guinea pig might be uniquely sensitive to respiratory irritants, results from this species were not used. Data from the rat and dog indicate a LOAEL for respiratory distress at an exposure concentration of approximately 1,200 mg/m^3 for 1 to 1.5 hr (Table 4-3). Using an uncertainty factor of 10 to extrapolate from a LOAEL to a NOAEL and an additional uncertainty factor of 10 to extrapolate from animals to humans, the subcommittee developed a 1-hr EEGL of 10 mg/m^3 (i.e., 12 rounded to one significant digit). Assuming that Haber's law applies over relatively short exposure durations (i.e., 15 min to 6 hr), the corresponding 15-min and 6-hr EEGLs (rounded to one significant digit) are 40 and 2 mg/m^3, respectively. Therefore, the subcommittee recommends 40, 10, and 2 mg/m^3 for 15 min, 1 hr, and 6 hr, respectively, for the EEGLs based on the animal data.

To check their recommendations against the available human information, the subcommittee also estimated an EEGL based on the ACGIH (1991) report. Using a divisor of 10 to extrapolate from a LOAEL of 100 mg/m^3 in humans to a NOAEL and assuming the concentration is "intolerable" in an exposure of approximately 1 hr, a 1-hr EEGL is 10 mg/m^3. Again applying Haber's law, the corresponding 15-min and 6-hr EEGLs are 40 and 2 mg/m^3, respectively. Thus, the EEGLs that the subcommittee developed on the basis of the animal studies are consistent with the human information.

The maximum anticipated total dose for field exposures (i.e., the peak concentration of a single volley of L8A1 grenades) has been estimated to be about 500 mg/m^3, and such a cloud would be expected to persist for 1 to 3 min (Weimer et al. 1980). Irritation might be expected following exposure to phosphorus smoke condensates because of the high phosphoric acid content.

Permissible Exposure Guidance Levels (PEGL)2

The PEGL should be similar to the ACGIH TLV-TWA, which appears to protect the worker from occupational exposure. The PEGL is designed to protect specific populations (i.e., military personnel and munitions workers). The combustion products associated with exposure to RP-BR smoke exposure have been characterized chemically. Because the product of concern is primarily phosphoric acid, the existing TLV-TWA for phosphoric acid exposure seems appropriate for military personnel as well. Therefore, establishing another set of exposure limits is not necessary. The PEGL for an exposure of 8 hr per day for 5 days per week (i.e., 40 hr per week) is 1.0 mg/m^3.

Public Exposures

Short-Term Public Emergency Guidance Level (SPEGL)3

Assuming that the general population comprises a wide variety of possibly sensitive individuals, an additional uncertainty factor of 10 is appropriate to extrapolate from an EEGL to a level

[2] Guidance for repeated exposure of military personnel during training exercises.
[3] Guidance for a rare, emergency situation potentially resulting in an exposure of the public to a military-training smoke.

protective of the general public. Thus, the SPEGLs for a single emergency exposure for RP-BR smoke are 4.0, 1.0, and 0.2 mg/m^3 for exposure durations of 15 min, 1 hr, and 6 hr, respectively (i.e., the corresponding EEGL values divided by 10).

Permissible Public Exposure Guidance Level (PPEGL)4

The possibility of repeated contamination of the air during military operations creates the need for establishing some guidance level for the communities in close proximity to Army operations. The Army has estimated that nearby community exposures to smokes and obscurant exposures might be as long as 8 hr per day, perhaps for a lifetime.

The TLV-TWA that has been set for phosphoric acid provides an acceptable concentration to which nearly all workers might be repeatedly exposed, day after day, without adverse effects during their working lifetime. To extend this exposure level to the general population requires the incorporation of an additional uncertainty factor of 10, because the general population might include individuals who are more sensitive to such exposures than are healthy workers.

Dividing the existing TLV-TWA for phosphoric acid of 1.0 mg/m^3 by a factor of 10 (for sensitive subpopulations), the PPEGL would be 0.1 mg/m^3.

Summary of Subcommittee Recommendations

The subcommittee's recommendations for exposure limits for RP-BR smoke for military personnel are summarized in Table 4-5. The recommendations for RP-BR-smoke concentrations at the boundaries of military-training facilities are summarized in Table 4-6.

[4] Guidance for repeated exposures of public communities near military-training facilities.

TABLE 4-5 EEGLs and PEGL for RP-BR Smoke for Military Personnel

Exposure Guideline	Exposure Duration	Guidance Level (mg/m^3)
EEGL	15 min	40
	1 hr	10
	6 hr	2
PEGL	8 hr/d, 5 d/wk	1

TABLE 4-6 SPEGLs and PPEGL for RP-BR Smoke at the Boundaries of Military-Training Facilities

Exposure Guideline	Exposure Duration	Guidance Level (mg/m^3)
SPEGL	15 min	4
	1 hr	1
	6 hr	0.2
PPEGL	8 hr/d, 5 d/wk	0.1

RESEARCH NEEDS

The subcommittee recognizes the need for further research to better understand the potential toxicity of RP-BR. Research in the following areas will provide better insight into possible health effects of inhalation of RP-BR and help to determine with greater confidence a guidance level that is not overly conservative but is scientifically defensible.

- Short-term (e.g., 10 min to 8 hr) inhalation studies are needed to evaluate the degree to which Haber's law applies to RP-BR smoke.
- Research should be undertaken to determine possible reproductive and developmental toxicity in mammals.
- Studies should be conducted to determine and to identify possible sensitive populations.

- Performance studies for military personnel operating in a smoke environment could determine if significantly impaired performance could occur.
- Documentation of the effects of exposure on humans should be developed to the extent possible.
- Pharmacokinetic and metabolism studies should be conducted to understand the mechanism of toxicity of RP-BR smokes.
- Further studies on possible mutagenic effects of RP-BR would be appropriate and would aid in clarifying whether this substance is a possible clastogen.

Finally, the subcommittee notes that Army personnel who work with this smoke, trainers in particular, are potentially a rich source of information on the health effects of the smoke. The subcommittee recommends that the Army conduct a prospective study with appropriate controls in which pulmonary-function tests and routine chemistry tests (panel 20 plus Mg and thyroid tests as a minimum requirement) are performed on personnel who are exposed repeatedly to the smoke.

REFERENCES

ACGIH (American Conference of Governmental Industrial Hygienists). 1991. Documentation of the Threshold Limit Values and Biological Exposure Indices, 6th Ed. American Conference of Governmental Industrial Hygienists, Cincinnati, Ohio.

Aranyi, C. 1983. Research and Development on Inhalation Toxicologic Evaluation of Red Phosphorus/Butyl Rubber Combustion Products, Phase 2 Report . AD-A158323. IIT Research Institute, Chicago.

Aranyi, C. 1984. Research and Development on Inhalation Toxicologic Evaluation of Red Phosphorus/Butyl Rubber Combustion Products, Phase 3 Report. AD-A173549. IIT Research Institute, Chicago.

Aranyi, C. 1986. Research and Development on Inhalation Toxicologic Evaluation of Red Phosphorus/Butyl Rubber Combustion Products,

Final (Phase 4) Report. AD-A189254. IIT Research Institute, Chicago.

Aranyi, C., M.C. Henry, S.C. Vana, R.D. Gibbons, and W.O. Iverson. 1988a. Effects of multiple intermittent inhalation exposure to red phosphorus/butyl rubber obscurant smokes in Sprague-Dawley rats. Inhalation Toxicol. Premier Issue:65-78.

Aranyi, C., S.C. Vana, J.N. Bradof, and R.L. Sherwood. 1988b. Effects of inhalation of red phosphorus/butyl rubber combustion products on alveolar macrophage responses in rats. J. Appl. Toxicol. 8:393-398.

Ballou, J.E. 1981. Chemical Characterization and Toxicologic Evaluation of Airborne Mixtures, Final Report. AD-A102678. Pacific Northwest Laboratories, Richland, Wash.

Brazell, R.S., J.H. Moneyhun, and R.W. Holmberg. 1984. Chemical Characterization and Toxicological Evaluation of Airborne Mixtures. Chemical and Physical Characterization of Phosphorus Smokes for Inhalation Exposure and Toxicology Studies. Final Report. ORNL/TM-9571. AD-A153 824. Oak Ridge National Laboratory, Oak Ridge, Tenn.

Lundy, D., and J. Eaton. 1994. Occupational Health Hazards Posed by Inventory U.S. Army Smoke/ Obscurant Munitions (Review Update). WRAIR/RT-94-0001. AD-A276 774. Walter Reed Army Institute of Research, Washington, D.C.

Mitchell, W.R., and E.P. Burrows. 1990. Assessment of Red Phosphorus in the Environment. Tech Rep. 9005. AD-A221704. U.S. Army Biomedical Research and Development Laboratory, Frederick, Md.

Ramsey R.S., J.H. Moneyhun, and R.W. Holmberg. 1985. The Chemical and Physical Characterization of XM819 Red Phosphorus Formulation and the Aerosol Produced by Its Combustion, Final Report. ORNL/TM-9941. Oak Ridge National Laboratory, Oak Ridge, Tenn.

Shinn, J.H., S.A. Martins, P.L. Cederwall, and L.B. Gratt. 1985. Smokes and Obscurants: A Health and Environmental Effects Data Base Assessment, Phase 1 Report. AD-A185377. Lawrence Livermore National Laboratory, Livermore, Calif.

Weimer, J.T., G. Affleck, J. Preston, J. Lucey, J. Manthei, and F. Lee. 1977. The Acute Effects of Single Exposure to United Kingdom Red Phosphorus Screening Smoke in Rats , Guinea Pigs, Rabbits, and Dogs. Tech. Rep. ARCSL-TR-77052. Chemical Systems Laboratory, U.S. Army Armament, Munitions and Chemical Command, Aberdeen Proving Ground, Edgewood, Md.

Weimer, J.T., G.E. Affleck, R.L. Farrand, F.K. Lee, and R.J. Pellerin. 1980. The Acute and Chronic Effects of Repeated Exposure to United Kingdom Red Phosphorus Screening Smokes in Rats, Mice, Guinea Pigs, and Rabbits. Tech. Rep. ARCSL-TR-79053. Chemical Systems Laboratory, U.S. Army Armament, Munitions and Chemical Command, Aberdeen Proving Ground, Edgewood, Md.

5

Hexachloroethane Smoke

BACKGROUND INFORMATION

The toxicity of hexachloroethane (HCE) smoke (referred to as HC smoke)[1] is attributed to the production of zinc chloride ($ZnCl_2$). Karlsson et al. (1991) compared acute inhalation exposures of HC smoke generated with zinc oxide (ZnO) with those generated with titanium dioxide (TiO_2). TiO_2-HCE smoke proved to be far less toxic than ZnO-HCE smoke, and ZnO-HCE smoke was lethal, causing gross pathological pulmonary injuries and death due to pulmonary edema. These authors also compared exposures to titanium tetrachloride ($TiCl_4$) gas with exposures to $ZnCl_2$ aerosol. No animals died from exposure to $TiCl_4$ at concentrations up to 2,900 mg/m^3 for 10 min, whereas the LC_{50} for $ZnCl_2$ was 2,000 mg/m^3 for a 10-min exposure.

Most reports of accidental human exposures to HC smoke indicate symptoms consistent with exposures to the $ZnCl_2$ component released when the smoke bomb is ignited. Therefore, the exposure-response assessments for HC smoke are probably most reliably interpreted, given present data, on the basis of the exposure-response data for $ZnCl_2$.

---

[1] In this chapter, HCE refers to the compound hexachloroethane, and HC smoke is the term used by the military for smoke produced by combusting HCE with zinc oxide and producing zinc chloride.

Military Applications

HC smoke is used by the U.S. military in a wide variety of munitions, some of which are shown in Table 5-1. HC smoke is produced by burning a mixture containing roughly equal parts of HCE and ZnO and approximately 6% granular aluminum.

Combustion Products

The smoke mixture in a smoke bomb or grenade is initially ignited by a pyrotechnic starter mixture. The reaction is self-perpetuating and exothermic. The overall reaction was summarized by Cichowicz (1983):

$$2\ Al + C_2Cl_6 + 3\ ZnO \rightarrow 3\ ZnCl_2 + Al_2O_3 + 2\ C + heat.$$

Another reaction produces carbon monoxide instead of solid carbon. $ZnCl_2$ leaves the reaction zone as a hot vapor. On cooling below the condensation point, it nucleates to form an aerosol that rapidly absorbs water from the surrounding atmosphere. Hydrated $ZnCl_2$ particles then scatter light, thereby obscuring vision. Because of $ZnCl_2$'s affinity for water, the aerosol likely consists of the hydrated forms of $ZnCl_2$ under most atmospheric conditions (Katz et al. 1980). A starter mixture containing silicon, potassium nitrate, charcoal, iron oxide, granular aluminum, cellulose nitrate, and acetone, which is required to initiate the reaction, might generate very small amounts of other airborne contaminants. However, the acute toxic effects of exposure to HC smoke are considered to arise primarily from inhalation of the $ZnCl_2$ component, which comprises almost two thirds of the total mass of HC smoke (Table 5-2). All measurements of HC smoke are expressed in this chapter as milligrams of $ZnCl_2$, unless noted otherwise.

The munitions listed in Table 5-1 all use slightly different chemical mixtures (Novak et al. 1987). An analysis of trace materials

TABLE 5-1 Characteristics of HC Smoke Munitions

Smoke-Pot Munitions[a]	Container Size (in.)	Filling Weight (lb)	Ignition Method	Weight (lb) (approx.) with Fuse	Delay Time (sec)	Burning Time (min)
Smoke pot, HCE, 10-lb, M1	9 by 5.5 diameter	10	Matchhead and scratcher block or electrical	12.5	10	5-8
Smoke pot, HCE, 30-lb, ABC-M5	9.5 by 8.5 diameter	31	Matchhead and scratcher block or electrical	33	20-30	2-22
Smoke pot, floating, HCE, M4A2	13 by 12 diameter	27.5	M207A1 smoke-pot fuse	38	10-20	10-15
Smoke grenade, HCE, M8	4.75 by 2.5 diameter	1.2	M201A1 fuse	1.5	0.7-2[b]	1.7-2.5
Cartridge,[c] 105-mm, HCE, M84A1		12.3	Mechanical, time, and super-quick fuse	13.0	60-90	3
Projectile,[d] 155-mm, HCE, M116A1		25.8	Mechanical, time, and super-quick fuse	26.2	60-90	4

[a] All HC smokes are type C, which contains granular aluminum, hexachloroethane, and zinc oxide. Other types of HC smoke were used in the early years of smoke generation.

[b] Time to functioning after release of safety lever.

[c] No future production for the M84A1 was planned as of 1983.

[d] M116A1 was completing its production life cycle in 1983 and would be replaced by XM 825 white phosphorus fill.

Source: Cichowicz (1983).

in HC smoke mixtures found common zinc impurities (Katz et al., 1980). Arsenic ranged from 0.13 to 5.0 microgram per gram (μg/g), mercury from 0.35 to 0.60 μg/g, cadmium from 53 to 1,523 μg/g, and lead from 50 to 858 μg/g. The cadmium and lead concentrations displayed a strong negative correlation.

TABLE 5-2 Approximate Composition of HC Smokea

Constituent	Estimated Mass Fraction, %
Zinc chloride	62.5
Zinc oxide	9.6
Iron oxide[b]	10.7
Aluminum oxide[b]	5.4
Lead oxide[b]	1.0
---------------------------- ----------------	---------------------------- ----------------
Total particulate phase	89.2
Chlorinated vapors	10.8

[a] The analysis does not take into account any liquid water that associates with $ZnCl_2$.
[b] These metals were assumed to be present as the oxide for purposes of calculating the mass fraction.
Source: DeVaull et al. (1989)

Trace gas-phase products were measured in a field test of a standard M5-HCE 30-lb smoke pot (Katz et al. 1980). Table 5-3 shows the resulting gas-phase products at two distances from the pot. Laboratory tests showed that hydrogen chloride (HCl) vapor formation decreased with increasing relative humidity (Katz et al. 1980). Because the field test was performed at -2°C, humidity was probably low. Thus, HCl vapor concentrations shown in Table 5-3 could be much higher than those produced under more humid conditions. However, Katz et al. (1980) speculated that under humid conditions, HCl is absorbed from the vapor phase into $ZnCl_2$ and water aerosol particles. Therefore, with increasing humidity, exposure of respiratory tissue to HCl might shift to lower portions of the lung, because small aerosol particles can penetrate to the lower lung and vapor can be removed readily from incoming air in the upper airways. During four field tests, estimated chlorine (Cl_2) production ranged from 3 to 19 mg/g of mixture combusted.

Aerosol formation was studied in a chamber using scaled-down HC smoke pots (Katz et al. 1980). The mass-median and the count-mean diameters produced were approximately 0.4 μm and 0.3 μm, respectively, averaged over 29 experiments. The observed size distribution was log-normal at lower initial particle concentrations (83×10^6 particles per cubic centimeter) and multimodal at higher concentrations. Relative humidity had no consistent effect on the total particulate concentration or the particle size. As the aerosols aged over a 2-hr period, the mass-median and count-mean diameters nearly doubled as the particle concentration decreased by a factor of about 6.

Laboratory-produced HC smoke consisted primarily of Zn^{+2} and Cl^{-1} (Katz et al. 1980). The aluminum content ranged from 0.49% to 4.06% of the Zn content, with a mean of 1.79%. The lead content ranged from 0.13 to 2.2 μg/mg of Zn, and the cadmium content ranged from 0.18 to 5.0 μg/mg of Zn. The ratios of both lead and cadmium to zinc were slightly higher than the ratios in the unburned mixture and were well correlated with them.

Physical and Chemical Properties of Zinc Chloride

CAS no.:	7646-85-7
Molecular formula:	$ZnCl_2$
Molecular weight:	136.29
Chemical name:	Zinc chloride
Synonyms:	Butter of zinc, zinc butter, zinc
Physical state:	Solid
Melting point:	290°C
Boiling point:	732°C
Density:	2.907 at 25°C
Vapor pressure:	1 mm Hg at 428°C
Solubility:	4.32×10^6 mg/L at 25°C 6.15×10^6 mg/L at 100°C 1 g/1.3 mL ethyl alcohol 1 g/2 mL glycerol 1 g/0.25 mL 2% hydrochloroacetic acid

TABLE 5-3 Chemical Analysis of Vapor Reaction Products from Field Test of 30-lb Military HC Smoke Pot

Distance from Mount to Pot (cm)	CO (ppm)	HCl (ppm)	$COCl_2$ (ppm)	CCl_4 (ppm)	C_2Cl_4 (ppm)	C_2Cl_6 (ppm)	C_6Cl_6 (ppm)
= 15	<1	1128	30	33	36	nd	nd
= 15	<1	1958	16	8	9	nd	nd
= 15	<1	5693	30	57	192	40	103
= 15	<1	6822	20	36	81	40	95
= 200	<1	1137	1	1	2	nd	nd

Abbreviation: nd, not determined.
Source: Katz et al. (1980).

Occurrence and Use

$ZnCl_2$ is used in preserving wood and in the manufacture and dyeing of fabrics. In addition to its use in military obscurants, $ZnCl_2$ is also the major ingredient in smoke from smoke bombs used for crowd dispersal and in fire-fighting exercises (by both military and civilian communities) (ASTDR 1994). $ZnCl_2$ also has uses in dental, medical, and household applications, as well as in herbicides (ATSDR 1994).

Military Exposures

Inhalation is expected to be the most important route of exposure. Undoubtedly excessive exposure has occurred in the military. Hill and Wasti (1978) summarized case reports of accidental exposures, many of which were fatal. The fatal exposures resulted from the discharge of HC smoke devices in enclosed spaces. The exposures in these reported cases generally are poorly characterized and represent only the most extreme conditions.

Very few data on HC smoke exposure are available for typical atmospheric conditions. Young et al. (1989) collected air samples during 1-hr demonstrations of M5 smoke pots and M8 smoke grenades at the U.S. Army Chemical School. Cadre members ignited these devices and students remained upwind. Personal and area samples were collected using mixed-cellulose-ester filters and high-flow personal-sampling pumps. Also, the particle-size distribution was characterized using cascade impactors. Zinc in the samples was measured by atomic absorption spectroscopy. Exposures to zinc for three cadre members were 0.0375, 0.0652, and 0.0776 mg/m^3 , or 0.0781, 0.1358, and 0.1616 mg/m^3 as $ZnCl_2$. The mass-median diameters of the particles ranged from 0.4 to 2.8 μm. Thus, a large portion of the particulate mass was respirable.

Simulated combat training during a military operation on urban terrain (MOUT) exercise indicated that trainees and instructors are exposed to $ZnCl_2$ in concentrations ranging from 0.02 to 0.98 mg/m^3 during a 225-min period. The average exposure concentration was 0.26 mg/m^3 with a standard deviation of 0.26 mg/m^3 (or 59 mg•min/m^3 over a 225-min exposure) (Young 1992, as cited in Lundy and Eaton 1994).

The most extensive field study of HC smoke, reported by DeVaull et al. (1989), shows the average smoke composition observed over five experiments. Composition and sampling location varied from test to test. In these tests, the total weight of smoke released ranged from 218.5 to 229.3 kg for groups of 18 to 20 M5 smoke pots. The particle mass-median diameters ranged from 0.77 to 1.05 μm, with geometric standard deviations from 1.78 to 2.36. Those particle sizes generally agree with those found by Katz et al. (1980) for aged aerosol and by Young et al. (1989), confirming that the aerosol has a large respirable mass fraction. DeVaull et al. (1989) also measured four specific chlorinated organic compounds. The geometric mean ratios of tetrachloromethane, tetrachloroethylene, hexachloroethane, and hexachlorobenzene to zinc in HC smoke were found to be 0.014, 0.009, 0.010, and 0.030, respectively.

A computer simulation of exposure to smoke released from 41 M5 smoke pots was carried out for a wind speed of 6 m/sec using the HAZARD2 program (Cichowicz 1983). The roughly rectangular area with exposures expected to exceed 60,000 mg•min/m^3 was 1,400 m wide in the cross-wind direction and about 1,000 m downwind of the release. Donohue et al. (1992) estimated that acute exposures in excess of 50,000 mg•min/m^3 can cause death or severe injury.

Table 5-4 shows calculations of Cichowicz (1983), who estimated minimum downwind distances from an M5 smoke pot necessary to limit exposures to designated concentrations and concentration-time profiles under various atmospheric conditions. The minimum distances downwind from one M5 smoke pot are listed in column 4 of Table 5-4. Similarly, distances are shown that are calculated to yield the concentration × time (CT) products ranging from the STEL (2 mg/m^3 × 15 min = 30 mg•min/m^3) to the highest estimated exposure level of 4,800 mg•min/m^3.

TABLE 5-4 Downward Distances from One M5 Smoke Pot (HC) Necessary to Limit Exposure to Designated ZnCl2 Concentrations and Concentration × Time Products

Atmospheric Condition	Wind Speed (m/sec)	Plume Rise (m)	Distance at Peak Concentration W 2 mg/ m^3 (m)	Distance at CT W 30 mg•min/ m^3 (m)	Distance at CT W 2,000 mg•min/ m^3 (m)	Distance at CT W 4,800 mg•min/ m^3 (m)
Night (very stable)	1	0	3,900a	3,700	190	100
	3	0	2,000	1,720	80	40
	1	18	3,600	3,200	–	–
Day (neutral stability)	1	0	1,100	950	80	50

Abbreviation: CT, concentration × time product.
Source: Adapted from Cichowicz (1983).

The potential for exposure during HC smoke training was explored in a cancer risk-assessment study (Novak et al. 1987). On the basis of the Army's Training Ammunition Management Information System (TAMIS), 15 bases using HC smoke munitions for training were identified. Fort Irwin in California was believed to have the greatest potential for exposure of all the sites in the United States and thus was chosen for study. At that base, Army forces train by mock combat with a simulated opponent force (OPFOR). A typical scenario is that OPFOR attacks the friendly force under cover of obscurants, often a combination of HC smoke and fog oil.

Three exposure categories were identified: smoke-generator squads, OPFOR, and friendly forces (Novak et al. 1987). The smoke-generator squads must stay 50 to 75 m directly downwind of the smoke pots. These squads were generating smoke 2 weeks per month for 10 months in 1982 and 3 weeks per month for 12 months in 1986. They have the greatest consistent exposure potential. Friendly forces rotate through this training and seldom train for more than 2 to 3 months per year. OPFOR attacks one friendly force after another over the course of a year and spends much of its training time under smoke cover; the prescribed OPFOR tactics in the 1980s called for heavy use of obscurants.

Novak et al. (1987) estimated the "worst-practical-case" long-term exposure at Fort Irwin for a soldier deploying smoke pots and standing 50 m directly downwind of the smoke pot. On the basis of usage of HC-smoke munitions in fiscal year 1982 and an assumed 2-year tour of duty, each person in the smoke-generator squad was considered exposed to emissions from 262 M5 smoke pots. Using air-dispersion modeling under very stable conditions and a wind speed of 2 m/sec, Novak et al. (1987) estimated that each member of the smoke generator squad was exposed to HC smoke at a total concentration of 9,916 mg•min/m^3. Again assuming a 2-year tour of duty, exposure concentrations were estimated to range from a minimum of 52.2 to nearly 20,000 mg•min/m^3 for the OPFOR, from 0.04 to 0.48 mg•min/m^3 for the friendly force, and approximately 62 mg•min/m^3 total HC smoke for the nearby community.

In addition to exposure to the smoke itself, workers manufacturing HC smoke munitions might be exposed to toxic materials. The most significant possible exposure is to HCE. HCE is a white crystalline solid with a vapor pressure equivalent to 770 ppm at 25°C (Eaton et al. 1994). Its camphorlike odor can be detected as low as 0.15 ppm. The HCE concentration in HC-smoke munitions production areas was found to be above the American Conference of Governmental Industrial Hygienists (ACGIH) Threshold Limit Value (TLV) of 9.7 mg/m^3 (Selden et al. 1993). Workers in these areas are protected with either air-supplied hoods at fixed work stations or full-face-piece respirators with both filters and organic vapor cartridges. Nevertheless, the plasma concentrations of HCE in workers rose from 0.08 ± 0.14 μg/L to 7.30 ± 6.04 μg/L after working for more than 5 weeks in the loading and packing operations (Selden et al. 1993).

TOXICOKINETICS

The toxicokinetics of ingested Zn has been studied in humans and animals, but inhaled Zn has not been systematically investigated (Donohue et al. 1992; ATSDR 1994). Intestinal absorption of radiolabeled $ZnCl_2$ increased up to 55% at low concentrations of Zn (90 μmol or less) in humans and declined with much higher concentrations of Zn (Payton et al. 1982). Uptake can be increased by dietary zinc deficiency (Istfan et al. 1983). Data on distribution and excretion of inhaled $ZnCl_2$ are not available, but at least one report of unmasked humans who succumbed following acute exposure to a high concentration of HC smoke showed increased concentrations of Zn in the lung and striated muscle (Hjortso et al. 1988).

TOXICITY SUMMARY

Effects in Humans

$ZnCl_2$ is corrosive and astringent and known to cause burning of moist body surfaces, including the respiratory and gastrointestinal tract. It has been reported to damage nerve endings in the nasal passages and to cause eye burns, damaging smell and vision. The upper respiratory tract is most affected by exposure (ATSDR 1994).

Oral Exposures

Acute oral exposures of humans to $ZnCl_2$ are associated with numerous symptoms that include vomiting, diarrhea, lethargy, and irritation of the mouth, throat and stomach. Ingestion can produce corrosive gastritis and liver necrosis. In one case of food poisoning (83 mg of Zn per 100 g of apples), characteristic symptoms were salivation; edema of the glottis; difficulty swallowing and massive swelling of the lips; pain in the mouth, throat, and epigastrium; recurrent intense vomiting; severe abdominal pain; and bloody diarrhea. Concentrations of 225 to 450 mg are known to be emetic. Two incidents of mass food poisoning produced symptoms that included abdominal cramping and occasional nausea and vomiting (ATSDR 1994).

Inhalation Exposures

One-Time Exposures

All the effects of human exposure to HC smoke are attributed to the $ZnCl_2$ component of the smoke.

Lethality. Death can occur with exposures to $ZnCl_2$ and has been attributed to respiratory insufficiency due to edema of the lungs or acute respiratory distress syndrome. The lethal dose of $ZnCl_2$ in humans has been estimated to be 50,000 mg·min/m^3 (see Table 5-5), an exposure that could be achieved according to Cullumbine (1957) by one generator in a 100-ft ^{3} room within 2 to 3 min. Death is usually delayed by several days. For those who survive an acute exposure, recovery can be protracted.

Pulmonary Effects. Pulmonary effects include dyspnea, chest constriction, retrosternal and epigastric pain, hoarseness, cough, lacrimation, expectoration, and occasional hemoptysis. Sequelae can include cyanosis, elevated pulse, fever, and widespread edema. Exposure to low concentrations results in moderate presentation of these symptoms (Donohue et al. 1992). Cullumbine (1957) reported that concentrations of $ZnCl_2$ at 80 mg/m^3 for 2 min (160 mg·min/m^3) produced slight nausea and cough, and 120 mg/m^3 for 2 min (240 mg·min/m^3) resulted in irritation of the nose, throat, and chest; cough; and nausea. The lower limit of detection of HC smoke by humans is reported to be approximately 40 mg/m^3 (Schenker et al. 1981). In an accidental exposure case, pneumonitis was reported in teens after exposure to $ZnCl_2$ smoke produced by a grenade at a concentration of 4,075 mg/m^3 (Johnson and Stonehill 1961). The exposed individuals experienced nausea and other respiratory symptoms, such as those mentioned above. Table 5-5 summarizes the effects of inhaled $ZnCl_2$ smoke at various combinations of concentration and time.

Repeated Exposures

Virtually no data are available at present on the effects of repeated exposure of neurological, reproductive, developmental, or immunological effects of $ZnCl_2$ in humans.

TABLE 5-5 Effects of Inhaled ZnCl2

Concentration × Time (mg·min/m3)	Effect
< 160	Essentially no effect; some awareness of presence
160-240	Noticeable irritation of nose, throat, and chest
1,700-2,000	Marked irritation: Hospitalization and treatment required
20,000	Severe irritation; chemical pneumonia: Hospitalization and treatment required
50,000	Massive injury; fatality

Sources: Stocum and Hamilton (1976) as cited in Donohue et al. (1992).

Effects in Animals

Dermal and Ocular Exposures

$ZnCl_2$ has been shown to be a skin and eye irritant in animal studies. When applied to the shaved skin of guinea pigs, weight loss was evident. In rabbits, severe erythema was noted both on abraded and unabraded skin, and eye application induced severe corneal damage, an effect manifest 4 days to 2 weeks after application (ATSDR 1994).

Oral Exposures

One-Time Exposures

Domingo et al. (1988) reported an acute oral LD_{50} for $ZnCl_2$ of 528 mg/kg of body weight in rats and 605 mg/kg of body weight in mice.

Repeated Exposures

Hematological Effects. Rats fed $ZnCl_2$ in their food ad libitum for 7 days per week for 4 weeks were reported to have a decrease in hemoglobin to 85% of control values. The lowest-observed-adverse-effect level (LOAEL) reported for such effects was calculated to be a dose of Zn at 12 mg/kg per day, equivalent to $ZnCl_2$ at 25 mg/kg of body weight per day (Zaporowska and Wasilewski 1992).

Reproductive Effects. Rats fed 25 mg of Zn per day as $ZnCl_2$ for 8 weeks showed an increased frequency of sperm with altered chromatin structure (Evenson et al. 1993).

Inhalation Exposures

One-Time Exposures

Lethality. The acute LCT_{50} (product of concentration × time lethal to 50% of the test animals) for exposure to $ZnCl_2$ is reported to be 11,800 mg·min/m^3 for mice (Cullumbine 1957).

Pulmonary Effects. Brown et al. (1990) compared the effects of acute HC smoke inhalation (11,580 mg·min/m^3 as HC smoke or 4,900 mg·min/m^3 as $ZnCl_2$, assuming 20% of the total smoke is Zn and 42% is $ZnCl_2$; see Marrs et al. 1989) for 60 min to instillation exposures (2.5 mg/kg) in rats. All exposed animals experienced respiratory distress. Mortality was higher in the smoke-exposed animals; respiratory distress developed more slowly after instillation but lasted longer. Both groups showed edema of the lungs, destructive alveolitis, and macrophage infiltration, followed by development of fibrosis.

Repeated Exposures

Lethality and Pulmonary Effects. Marrs et al. (1988) determined the effects of repeated exposures to freshly generated ZnO and HC smoke in female mice, rats, and guinea pigs. Animals were exposed to three air concentrations of smoke (Zn at 0, 1.3, 12.8, or 122 mg/m^3) 1 hr per day, 5 days per week, for 20 weeks, for a total of 100 1-hr exposures.

The highest exposure concentration caused excessive mortality in both guinea pigs and mice. Inflammatory changes, such as edema, emphysema, and macrophage infiltration in the lungs of rats and guinea pigs, also occurred at the highest exposure concentration. The middleand low-exposure groups exhibited normal survival rates, and no adverse pulmonary effects were evident. The no-observed-adverse-effect level (NOAEL) was determined to be 12.8 mg of Zn per cubic meter, or 26.6 mg of $ZnCl_2$ per cubic meter, assuming that all the Zn in the smoke is in the form of $ZnCl_2$.

Other Systemic Effects. Marrs et al. (1988) reported no effects on other systems, including cardiovascular, gastrointestinal, hepatic, renal, immune, or reproductive systems. No effects on growth were observed at any exposure concentration.

Carcinogenic and Mutagenic Effects. The major histopathological finding of Marrs et al. (1988) was an increase in the incidence of alveologenic carcinomas in mice exposed to the highest air concentration (i.e., Zn at 122 mg/m^3, or $ZnCl_2$ at 254 mg/m^3, assuming that all Zn is in the form of $ZnCl_2$). International Agency for Research on Cancer (IARC) has not evaluated $ZnCl_2$ for its carcinogenicity. Genotoxicity studies in bacterial and mammalian-cell culture test systems provide no evidence that $ZnCl_2$ is mutagenic (ATSDR 1994).

Summary of Toxicity Data

Tables 5-6 and 5-7 summarize exposure-response information for the lethal and nonlethal effects, respectively, resulting from inhalation exposure to HC smoke or $ZnCl_2$ aerosol.

TABLE 5-6 Lethality Resulting from Inhalation Exposure to HC Smoke or ZnC12 Aerosol (expressed as milligrams of ZnC12)

Species	Exposure Frequency and Duration	Exposure Level	End Point and Comments	Reference
Human	One time	50,000 mg•min/m^3	Some lethality	Donohue et al. 1992
Rat	10 min	12,500 mg•min/m^3	0/3 died	Karlsson et al. 1991
		19,600	2/3 died	
		25,400	2/3 died	
		40,600	3/3 died	
Mouse	One time	11,800 mg•min/m^3	LCT_{50}	Cullumbine 1957
Mouse	60 min	4,900 mg•min/m^3	Excess mortality	Brown et al. 1990[a]
Mouse (female), rat, guinea pig	1 hr/d, 5 d/ wk, 20 wk	26.6 mg/; 254 mg/m^3	NOAEL; LOAEL	Marrs et al. 1988[b]

Abbreviations: hr, hour(s); d, day(s); wk, week(s); LCT_{50}, product of concentration × time that is lethal to 50% of the test animals.

[a] Assumes 42% of 11,580 mg•min/m^3 total HC smoke is $ZnCl_2$.

[b] Assumes all Zn present is in form of $ZnCl_2$; see text.

TABLE 5-7 Summary of Exposure-Response Data for Nonlethal Effects from Exposure to HC Smoke or ZnCl2 (expressed as milligrams of ZnCl2)

Category and Species	Exposure Frequency and Duration	NOAEL (mg/m^3)	LOAEL (mg/m^3)	End Point and Comments	Reference
Effects in Humans					
One-Time Inhalation Exposure					
Pulmonary Effects					
Experimental subjects	2 min	8160 mg•min/m^3	160 mg•min/m^3	2 min•80 mg/m^3 : Slight nausea and cough	Cullumbine 1957
Experimental subjects	2 min	—	240 mg•min/m^3	2 min•120 mg/m^3 : Irritation of nose, throat, chest; cough and nausea	Cullumbine 1957
Effects in Animals					
Repeated Oral Exposures					
Rat	7 d/wk, 4 wk	—	25 mg/kg-d	Hemoglobin decrease by 15%	Zaporowska and Wasilewski 1992
Rat	7 d/wk, 4 wk	—	52 mg/kg-d	Increased frequency of sperm with altered chromatin	Evenson et al. 1993
One-Time Inhalation Exposures					
Pulmonary Effects					
Mouse	60 min	—	4,900 mg•min/m^3	60 min•120 mg/m^3 : Respiratory distress and excess mortality	Brown et al. 1990

Repeated Inhalation Exposures					
Pulmonary Effects					
Mouse (female), rat, guinea pig	1 hr/d, 5 d/wk, 20 wk (100 hr total)	26.6 mg/m^3	254 mg/m^3	Inflammatory changes in the lungs; lethality	Marrs et al. 1988
Carcinogenic Effects					
Mouse (female), rat, guinea pig	1 hr/d, 5 d/wk, 20 wk (100 hr total)	26.6 mg/m^3	254 mg/m^3	Alveologenic carcinomas	Marrs et al. 1988

Abbreviations: hr, hour(s); min, minute(s); d, day(s); wk, week(s).

Noncancer Toxicity

The target organ for inhalation exposures to HC smoke in humans and animals is the respiratory tract. Data from humans indicate that a threshold for slight nausea and irritation of the nose, throat, and chest from exposure to $ZnCl_2$ is between 160 and 240 mg•min/m^3. At concentrations of 1,700 mg•min/m^3 and above, effects can be severe enough to require hospitalization and treatment. Data from animals are sparse but indicate a NOAEL for repeated 1-hr exposures to HC smoke with Zn at 12.8 mg/m^3 (or $ZnCl_2$ at 26.6 mg/m^3) in rodents and a LOAEL for inflammatory changes in the lung and death with Zn at 122 mg/m^3 (or $ZnCl_2$ at 254 mg/m^3), suggesting a relatively steep dose-response curve.

Example Noncancer Risk Assessment

According to reports of simulated combat training during a MOUT exercise, trainees and instructors are exposed to $ZnCl_2$ at concentrations ranging from 0.02 to 0.98 mg/m^3 for 225 min. The average exposure concentration was calculated to be 0.26 mg/m^3 with a standard deviation of 0.26 (Young 1992, as cited in Lundy and Eaton 1994). In the worst-case scenario, that concentration would result in exposures as high as 225 mg•min/m^3 (i.e., 1 mg/m^3 × 225 min). Thus, the current MOUT exercise might engender $ZnCl_2$ exposures that exceed the human threshold for adverse respiratory-tract effects at 160 to 240 mg•min/m^3. Additionally, the values from the simulated MOUT exercise do not specify distance from the source of the smoke pot or the direction of air movement with respect to the source and exposed personnel. As indicated in Table 5-4, moderate distances (i.e., 40 to 190 m) downwind of a smoke pot might result in exposure concentrations of 2,000 mg•min/m^3 or higher, which could result in the need for hospitalization and treatment (see Table 5-5).

The above example does not take into account the possibility that women of childbearing age, who might be pregnant, could be members of the exposed population. Virtually no human or animal data are available to provide any information on potential reproductive or developmental toxicity. In addition, no information is available on the effects of these exposures on nervous-system function. Combined with those caveats is the fact that the onset of toxicity tends to be delayed, and recovery, if it occurs, is slow.

Given the likely inability of the soldier to control distance and duration of exposure, the above example suggests that masks should be worn at all times during such exercises. Mask use is especially important because the irritant effects might degrade the performance of the soldiers.

Carcinogenicity

Although only one study of one species reported a positive carcinogenic response resulting from exposure to HC smoke (i.e., Marrs et al. 1988), the subcommittee decided to estimate the potential carcinogenic potency of HC smoke based on this study as a screen to determine whether further analysis is warranted. The derivation of the potential carcinogenic potency of HC smoke based on this one study is described below.

Marrs et al. (1988) exposed mice to HC smoke. The concentrations of Zn in the air were 0 (control), 1.3, 12.8, and 122 mg/m^3. Because Zn represented 20% of the smoke in these experiments, the concentrations of total smoke were 5 times higher: 0, 6.5, 64, and 610 mg/m^3. Assuming a 25-g mouse inhaled air at 0.0018 /hr during the 100 hr of exposure, each animal inhaled 0.18 of air containing HC smoke. The total doses of smoke for the groups were 0, 1.2 (i.e., 6.5 × 0.18), 12 (i.e., 64 × 0.18), and 110 mg (i.e., 610 × 0.18). Assuming a 2-year mouse lifetime, the average daily dose per kilogram of body weight is 0, 0.064, 0.63, and 6.0 mg/kg per day (i.e., total mg dose/body weight × 730 days).

This calculation assumes that the total dose was distributed evenly over the lifetime of the mouse and is used for estimating carcinogenic potency.

The incidences of alveologenic carcinoma at 18 months were 6 of 78 in the controls, 7 of 74 in the 1.3-mg/ group, 8 of 76 in the 12.8-mg/ group, and 15 of 50 in the 122-mg/ group (Marrs et al. 1988). Using the average daily doses calculated above and fitting a generalized multistage model to the cancer data provides an upper limit of the cancer risk (low-dose slope) of 0.036/mg/kg per day. This model was used for estimating the potency of HC smoke at low doses to provide a conservative estimate for screening purposes. The cancer potency is based on exposures expressed as milligrams per kilogram of body weight per day. Cancer potency based on scaling body weight to the 3/4 power would result in cancer risk estimates about a factor of 7 higher, or 0.25 mg/kg per day.

The subcommittee also estimated carcinogenic potency expressed in terms of $ZnCl_2$ instead of total HC smoke. Given that Zn comprised 20% of the HC smoke in the experiments of Marrs et al. (1988), $ZnCl_2$ should have comprised approximately 42% of the total smoke used in those experiments. If one assumes that essentially all the Zn present is in $ZnCl_2$ (although some will be present in an oxide form), the resulting upper 95% confidence limit of the cancer risk (low-dose slope) would be 0.086 mg/kg of body weight per day (i.e., the potency of total HC smoke of 0.036 mg/kg per day divided by 0.42). Cancer potency based on scaling body weight to the 3/4 power would be about a factor of 7 higher, or 0.6 mg/kg per day.

Lifetime cancer risks are generally based on the tumor incidence from a 2-year rodent bioassay. Studies of shorter duration might underestimate the true risk, because cancer incidence generally increases rapidly with age. On the other hand, animals were exposed to HC smoke during the first 20 weeks of the study. Because young animals might be more sensitive to HC smoke and have a longer time for tumors to develop from early-life exposure, the observed tumor incidence might be higher than would be observed if the same total dose were administered over 2 years. Because the tumor incidence could be adjusted upward and downward, no adjustment will be made based on the less-than-lifetime exposure duration.

Example Cancer Risk Assessment

Estimated Exposure

Novak et al. (1987) estimated exposures to HC smoke for smoke-generator operators. An operator was assumed to be 50 m downwind, and 100% of the available respirable dose was assumed to be absorbed. Under those conditions, the Army calculated the exposure concentration from a single smoke pot to be 1,120 mg•min/m^3 for an atmospheric stability condition defined by the Army as "neutral." Using an inhalation rate of 0.03 /min, the dose from a single smoke pot is 1,120 × 0.03 or 33.6 mg of HC smoke. An operator was assumed to be exposed to 262 releases during a 2-year tour of duty, resulting in 8,800 mg of total HC smoke inhaled (i.e., 262 × 33.6 mg). Averaged over a 70-year lifetime for a 70-kg person, the average daily dose is

$$8{,}800 \text{ mg}/(70 \text{ yr} \times 70 \text{ kg} \times 365 \text{ d/yr}) = 0.0049 \text{ mg/kg per day.}$$

Under "very stable" atmospheric conditions, the exposure concentrations from a single smoke pot is 9,900 mg•min/m^3, resulting in an average daily lifetime dose of 0.044 mg/kg per day for a smoke-pot operator. Note, however, that very stable and neutral atmospheric conditions are worst-case estimates (i.e., little or no wind to disperse the smoke plume by convection) and would not apply most of the time. Hence, assuming these conditions will overestimate risk. Average atmospheric stability conditions could be used instead to estimate risks for a particular facility.

In the community of Baker nearby Fort Irwin, Novak et al. (1987) estimated total lifetime exposure to be 62.3 mg•min/m^3.

Assuming an inhalation rate of 0.03 /min and 100% absorption, the total absorbed dose is 1.86 mg of HC smoke (i.e., 62.3 × 0.03). The average daily dose over a 70-year lifetime for a 70-kg person is

$$1.86\ \text{mg}/(70\ \text{yr} \times 70\ \text{kg} \times 365\ \text{d/yr}) = 1 \times 10^{-6}\ \text{mg/kg per day.}$$

This dose is quite small compared with the dose estimated for smoke-pot operators.

Estimated Cancer Risk

Using the information above, the subcommittee estimated a lifetime cancer risk by multiplying the upper limit of the cancer potency of HC smoke (0.036/ mg/kg per day) by the dose averaged over a lifetime. For smoke-pot operators on a 2-year tour of duty, the lifetime cancer risk for an average daily dose of 0.0049 mg/kg per day was estimated to be for a neutral atmospheric condition. For a very stable atmospheric condition, the average daily lifetime dose was estimated to be 0.0436 mg/kg per day, giving a lifetime cancer risk of

$$\text{Risk} < 1.8 \times 10^{-4}\ (\text{i.e.,} < 0.036 \times 0.0049)$$

$$\text{Risk} < 1.6 \times 10^{-3}\ (\text{i.e.,} < 0.036 \times 0.0436).$$

If average atmospheric conditions are assumed, the estimated risks for smoke-pot operators would be lower.

In the community of Baker, even using the unrealistic worst-case atmospheric conditions, the average daily lifetime dose was estimated to be 1×10^{-6} mg/kg per day, resulting in a lifetime cancer risk of less than 3.6×10^{-8} (i.e., 0.036×10^{-6}) throughout the lifetime of Baker residents.

EXISTING RECOMMENDED EXPOSURE LIMITS

The ACGIH (1991) proposed a TLV time-weighted average (TWA) of 1 mg/m^3 for 8 hr per day, 5 days per week (or 40 hr per week) and a 15-min TLV short-term exposure limit (STEL) of 2 mg/m^3 for $ZnCl_2$.

SUBCOMMITTEE EVALUATION AND RECOMMENDATIONS

On the basis of the toxicity information presented for $ZnCl_2$, the subcommittee developed exposure guidance levels for military personnel exposed during an emergency release or during regular-training exercises and for nearby communities to protect them from emergency or repeated releases of HC smoke.

Military Exposures

Emergency Exposure Guidance Level (EEGL)2

To define an EEGL for $ZnCl_2$, several factors must be taken into consideration. Based on reports from acute human inhalation exposures, CT products of 160 mg•min/m^3 produced slight nausea and cough, and 240 mg•min/m^3 was associated with irritation of the nose, throat, and chest and with nausea and cough (Cullumbine 1957). Donohue et al. (1992) summarized these data by suggesting that the noticeable irritation range for nose, throat, and chest is 160 to 240 mg•min/m^3.

The NOAEL of 160 mg•min/m^3 for $ZnCl_2$ is based on short-term exposures of humans and therefore can be used for the EEGL

[2] Guidance for a rare, emergency situation resulting in an exposure of military personnel.

without adjustment to account for extrapolation uncertainties (e.g., animal to human, LOAEL to NOAEL). Thus, for a 60-min exposure period, the EEGL would be 2.7 mg/m^3, rounded off to 3.0 mg/m^3. Similarly, for a 15-min exposure period, the EEGL would be 10 mg/m^3, and for a 6-hr exposure period, the EEGL would be 0.4 mg/m^3. The 15- and 60-min EEGLs exceed the current STEL value set by ACGIH (2 mg/m^3) but can be justified by the fact that the EEGL is set for a one-time emergency exposure situation which is not expected to reoccur.

Permissible Exposure Guidance Levels (PEGL)[3]

A PEGL is required because of the chronic intermittent exposures to smokes that are experienced by soldiers, which would approximate 50 8-hr exposures during a 2-year tour of duty. Virtually no human data are available upon which to base a PEGL. One repeated inhalation toxicity exposure study carried out by Marrs et al. (1988) identified a NOAEL for rodents as $ZnCl_2$ at 26.6 mg/m^3 for 1 hr per day, 5 days per week, for 20 weeks. Using that NOAEL, a divisor of 10 for uncertainties associated with sparse animal data and with the shorter daily duration of exposure (1-hr exposures for the rodent compared with 8-hr exposures for military personnel), and another divisor of 10 to extrapolate from animals to humans, the subcommittee recommends a PEGL for $ZnCl_2$ of 0.2 mg/m^3.

Comparison with Other Exposure Guidance Levels

The ACGIH 8-hr TLV-TWA for $ZnCl_2$ of 1.0 mg/m^3 is higher than both the 6-hr EEGL of 0.4 and the PEGL of 0.2 mg/m^3 recommended above. The ACGIH TLV-TWA is based on

[3] Guidance for repeated exposure of military personnel during training exercises.

unpublished reports that a 30-min exposure at 4.8 mg/m^3 caused mild, transient respiratory irritation and that 0.4 mg/m^3 (duration of exposure not specified) was not considered irritating. ACGIH did not indicate why Haber's law was not applied to the 30-min exposure level to extrapolate to longer exposure periods. Without evidence to the contrary, the subcommittee assumes that Haber's law is applicable and therefore recommends the lower values for the 6-hr EEGL and PEGL rather than the ACGIH recommendation for its 8-hr TLV-TWA.

The actual exposures that can be experienced if cancer risks are not to exceed 1×10^{-4}, for example, would require that a total dose of no more than 5,000 mg of HC smoke be experienced (i.e., the 8,800-mg HC smoke associated with a 1.8×10^{-4} cancer risk divided by 1.8 to yield a 1×10^{-4} cancer risk). Assuming 262 exposures during a 2-year tour of duty, the total dose from a single smoke pot could not exceed 19 mg (5,000 mg of HC smoke divided by 262 exposures). Assuming further an inhalation rate of 0.03/min, the exposure concentration generated by the smoke pot could not exceed 630 mg•min/m^3 (i.e., 19 mg divided by 0.03 /min) or approximately 10 mg•hr/m^3 (i.e., 630 mg•min/m^3 divided by 60 min/hr) for a total of 262 such exposures. In the case of the training staff who endure a greater number and duration of such exposures, the values would have to be altered accordingly.

Public Exposures

Short-Term Public Emergency Guidance Level (SPEGL)4

In calculating the SPEGL, the EEGL is divided by an uncertainty factor of 10 to account for the susceptible subpopulations of the general public (e.g., the elderly, chronically ill, and children) that could conceivably be exposed. This calculation results in a

[4] Guidance for a rare, emergency situation potentially resulting in an exposure of the public to a military-training smoke.

SPEGL, expressed as a CT product, of 16 mg•min/m^3. The corresponding 15-min, 60-min, and 6-hr SPEGLs are 1, 0.3, and 0.04 mg/m^3, respectively.

Permissible Public Exposure Guidance Level (PPEGL)5

The general public living or working near military-training facilities could experience chronic intermittent exposures. For the PPEGL, the PEGL is divided by an uncertainty factor of 10 to account for sensitive populations (e.g., the elderly, chronically ill, and children). A PPEGL of 0.01 mg/m^3 is recommended.

Comparing the outcome of the example of cancer-risk estimates with the PEGL suggests that, because the cancer risks for the nearby community, as computed above, are well below the usual level of concern, the PPEGL guidelines should be used.

Comparison of Recommendations with Conservative Screening Cancer-Risk Estimates

Given the one finding of cancer in mice exposed to HC smoke (Marrs et al. 1988), the subcommittee estimated a conservative cancer potency factor for HC smoke (see Carcinogenic Effects) and estimated the corresponding cancer risk associated with the recommended exposure guidance levels for military personnel and the public.

For a 60-min EEGL of 3 mg/m^3 for $ZnCl_2$, the total exposure of 3 mg averaged over a 75-hr lifetime for a 70-kg adult is 3 mg/m^3 (75 yr × 365 days × 70 kg) = 1.6×10^{-6} mg/kg per day. The cancer risk is estimated to be below $0.086 \times (1.6 \times 10^{-6}) = 1 \times 10^{-7}$.

For 400 hr of exposure of military personnel during a tour of duty at a PEGL of 0.1 mg/m^3, the total exposure would be 40 mg

[5] Guidance for repeated exposures of public communities near military-training facilities.

of $ZnCl_2$. For this exposure, the cancer risk is estimated to be less than 2×10^{-6}.

For the SPEGL, the possible lifetime exposure would be 0.3 mg of $ZnCl_2$, which would correspond to a cancer risk of less than 2×10^{-8}.

For 30 hr of exposure per year at a PPEGL of 0.01 mg/m^3 for $ZnCl_2$, the total annual dose would be 0.3 mg of $ZnCl_2$, corresponding to an annual cancer-risk estimate of less than 2×10^{-8}. For a 30-year period of residence in the community, the corresponding risk estimate would be less than 6×10^{-7}.

Thus, the subcommittee concludes that the exposure guidance levels recommended for military personnel and the public developed on the basis of noncancer end points are sufficiently low to represent a negligible (i.e., approximately 1×10^{-6} or less) cancer risk if the substance is a human carcinogen. The data available to date, however, are insufficient to conclude that $ZnCl_2$ is a human carcinogen.

Summary of Subcommittee Recommendations

Table 5-8 summarizes the subcommittee's recommendations for EEGLs and the PEGL for military personnel exposed to HC smoke. Table 5-9 summarizes the subcommittee's recommendations for SPEGLs and the PPEGL for military-training facilities to ensure that nearby communities are not exposed at concentrations that might cause adverse effects.

TABLE 5-8 EEGLs and PEGL for HC Smoke for Military Personnel

Exposure Guideline	Exposure Duration	Guidance Level (mg/m^3)[a]
EEGL	15 min	10
	1 hr	3
	6 hr	0.4
PEGL	8 hr/d	0.2

[a] Expressed in milligrams of $ZnCl_2$ per cubic meter.

TABLE 5-9 SPEGLs and PPEGL for HC Smoke at the Boundaries of Military Training Facilities

Exposure Guideline	Exposure Duration	Guidance Level (mg/m^3)[a]
SPEGL	15 min	1
	1 hr	0.3
	6 hr	0.04
PPEGL	8 hr/d	0.02

[a] Expressed in milligrams of $ZnCl_2$ per cubic meter.

RESEARCH NEED

It is clear from the data reviewed in this chapter that insufficient information is available to evaluate potential long-term toxicity of $ZnCl_2$. As noted above, almost all studies have focused on pulmonary effects, and little information exists on potential toxicity to other systems, such as the developmental or reproductive systems or nervous system. In addition, little information exists on effects resulting from repeated exposures and on the reversibility of observed effects. If use of HC smoke devices based on reactions of HCE with ZnO continues, then additional information with respect to effects on other organs and systems is required to ensure the health of military personnel and to prevent releases beyond military facilities that might pose risks to the general public.

REFERENCES

ACGIH (American Conference of Governmental Industrial Hygienists). 1991. Documentation of the Threshold Limit Values and Biological Exposure Indices, 6th Ed. American Conference of Governmental Industrial Hygienists, Cincinnati, Ohio.

ATSDR (Agency for Toxic Substances and Disease Registry). 1994. Toxicological Profile for Zinc (Update). TR-93/15. Agency for Toxic Substances and Disease Registry, Public Health Service, U.S. Department of Health and Human Services, Atlanta, Ga.

Brown, R.F.R., T.C. Marrs, P. Rice, and L.C. Masek. 1990. The histopathology of rat lung following exposure to zinc oxide/hexachloroethane smoke or instillation with zinc chloride followed by treatment with 70% oxygen. Environ. Health Perspect. 85:81-87.

Cichowicz, J.J. 1983. Environmental Assessment, Programmatic Life Cycle Environmental Assessment for Smoke/Obscurants. HC Smoke, Vol. 4. ARCSL-EA-83007. Chemical Research and Development Center, U.S. Army Armament, Munitions and Chemical Command, U.S. Army Aberdeen Proving Ground, Edgewood, Md.

Cullumbine, H. 1957. The toxicity of screening smokes. J. R. Army Med. Corps 103:119-122.

DeVaull, G.E., W.E. Dunn, J.C. Liljegren, and A.J. Policastro. 1989. Analysis Methods and Results of Hexachloroethane Smoke Dispersion Experiments Conducted as Part of Atterbury-87 Field Studies. AD-A216048. Prepared by Argonne National Laboratory, Argonne, Ill., for the U.S. Army Medical Research and Development Command, Fort Detrick, Frederick, Md.

Domingo, J.L., J.M. Llobet, J.L. Paternain, and J. Corbella. 1988. Acute zinc intoxication: Comparison of the antidotal efficacy of several chelating agents. Vet. Hum. Toxicol. 30:224-228.

Donohue, J.M., L. Gordon, C. Kirman, and W.C. Roberts. 1992. Zinc Chloride Health Advisory. Interagency Agreement (IAG) 85PP5869. Office of Water, U.S. Environmental Protection Agency, Washington, D.C., and the U.S. Army Medical Research and Development Command, Fort Detrick, Frederick, Md.

Eaton, J.C., R.J. LoPinto, and W.G. Palmer. 1994. Health Effects of Hexachloroethane (HC) Smoke. USABRDL-TR-9402. AD-A277 838. U.S. Army Biomedical Research and Development Laboratory, Fort Detrick, Frederick, Md.

Evenson, D.P., R.J. Emerick, L.K. Jost, H. Kayongo, and S.R. Stewart. 1993. Zinc-silicon interactions influencing sperm chromatin integrity and testicular cell development in the rat as measured by flow cytometry. J. Anim. Sci. 71:955-962.

Hill, H.G., and K. Wasti. 1978. A Literature Review–Problem Definition Studies on Selected Toxic Chemicals. Occupational Health and Safety and Environmental Aspects of Zinc Chloride, Vol. 5, Final Report. AD A056020. Franklin Institute Research Laboratories, Philadelphia.

Hjortso, E., J. Qvist, M.I. Bud, J.L. Thomsen, J.B. Andersen, F. Wiberg-Jorgensen, N.K. Jensen, R. Jones, L.M. Reid, and W.M. Zapol.

1988. ARDS after accidental inhalation of zinc chloride. Intensive Care Med. 14:17-24.

Istfan, N.W., M. Janghorbani, and V.R. Young. 1983. Absorption of stable ^{70}Zn in healthy young men in relation to zinc intake. Am. J. Clin. Nutr. 38:187-194.

Johnson, F.A., and R.B. Stonehill. 1961. Chemical pneumonitis from inhalation of zinc chloride. Dis. Chest 40:619-623.

Karlsson, N., I. Fangmark, I. Haggqvist, B. Karlsson, L. Rittfeldt, and H. Marchner. 1991. Mutagenicity testing of condensates of smoke from titanium dioxide/hexachloroethane and zinc/hexachloroethane pyrotechnic mixtures. Mutat. Res. 260:39-46.

Katz, S., A. Snelson, R. Farlow, R. Welker, and S. Mainer. 1980. Physical and Chemical Characterization of Fog Oil Smoke and Hexachloroethane Smoke. DAMD17-78-C-8085. AD-A080 936. IIT Research Institute, Chicago.

Lundy, D., and J. Eaton. 1994. Occutational Health Hazards Posed by Inventory U.S. Army Smoke/ Obscurant Munitions (Review Update). U.S. Army Medical Research Detachment, Wright-Patterson Air Force Base, Ohio.

Marrs, T.C., H.F. Colgrave, J.A.G. Edginton, R.F.R. Brown, and N.L. Cross. 1988. The repeated dose toxicity of a zinc oxide/hexachloroethane smoke. Arch. Toxicol. 62:123-132.

Novak, E.W., L.B. Lave, J.J. Stukel, and D.J. Schaeffer. 1987. A Revised Health Risk Assessment for the Use of Hexachloroethane Smoke on an Army Training Area. USA-CERL Tech. Rep. N-87/26. Construction Engineering Research Laboratory, U.S. Army Corps of Engineers, Champaign, Ill.

Payton, K.B., P.R. Flanagan, E.A. Stinson, D.P. Chodirker, M.J. Chamberlain, and L.S. Valberg. 1982. Technique for determination of human zinc absorption from measurement of radioactivity in a fecal sample or the body. Gastroenterology 83:1264-1270.

Schenker, M.B., F.E. Speizer, and J.O. Taylor. 1981. Acute upper respiratory symptoms resulting from exposure to zinc chloride aerosol. Environ. Res. 25:317-324.

Seldén, A., M. Nygren, A. Kvamlöf, K. Sundell, and O. Spångberg. 1993. Biological monitoring of hexachloroethane. Int. Arch. Environ. Health 65:S111-S114.

Stocum, W.E., and R.G. Hamilton. 1976. A Risk Analysis of Exposure to High Concentrations of Zinc Chloride Smoke. SAND76-0386. Sandia Laboratories, Albuquerque, N.Mex.

Young, J.Y., D.A. Smart, J.T. Allen, D.L. Parmer, A.B. Rosencrance, E.E. Brueggeman, and F.H. Broski. 1989. Field Exposure of Chemical School Students and Cadre to Fog Oil and Hexachloroethane (HC) Smokes. Tech. Rep. 8908. U.S. Army Biomedical Research and Development Laboratory, Fort Detrick, Frederick, Md.

Young, J.Y. 1992. Field Exposure of Infantry Soldiers to Hexachloroethane and Colored Smoke During a "Military Operation-Urban Terrain" Training. U.S. Army Biomedical Research and Development Laboratory, Fort Detrick, Frederick, Md.

Zaporowska, H., and E. Wasilewski. 1992. Combined effect of vanadium and zinc on certain selected hematological indices in rats. Comp. Biochem. Physiol. C 103:143-147.